Letts

GCSE IN A WEEK

PHYSICS

Caroline Reynolds

Revision Planner

Page	Day	Time (mins)	Title	Exam Board	Date	Time	Completed
Day 1							
4	1	15 mins	How Science Works	AEO			
6	1	15 mins	Motion	AEO			
8	1	15 mins	Calculating Motion	AEO			
10	1	15 mins	Graphs for Motion	AEO			
12	1	15 mins	When Forces Combine	AEO			
14	1	15 mins	Forces and Motion	AEO			
16	1	15 mins	Momentum and Stopping	AEO			
Day 2							
18	2	15 mins	Safe Driving	AEO			
20	2	15 mins	Energy and Work	AEO			
22	2	15 mins	Work and Power	AEO			
24	2	15 mins	Static Electricity	AEO			
26	2	15 mins	Electricity on the Move	AEO			
28	2	15 mins	Circuits	AEO			
Day 3							
30	3	15 mins	Electrical Power	AEO			
32	3	15 mins	Electricity at Home	AEO			
34	3	15 mins	Controlling Voltage	O			
36	3	15 mins	Electronic Applications	O			
38	3	15 mins	Producing Electricity	AEO			
40	3	15 mins	Motors and Transformers	AO			
42	3	15 mins	Electricity in the World	AEO			
Day 4							
44	4	15 mins	Harnessing Energy 1	AEO			
46	4	15 mins	Harnessing Energy 2	AEO			
48	4	15 mins	Keeping Warm	AEO			
50	4	15 mins	Heat	A (except Equations) O			
52	4	15 mins	Heat Calculations	AEO			
54	4	15 mins	Temperature	E			

How Science Works

Sometimes our opinions are based on our own prejudices; what we personally like or dislike.

Ideas

At other times, our opinions can be based on scientific evidence. Reliable and valid evidence can be used to back up our own opinions.

Variables

- An **independent** variable is the variable that we choose to change to see what happens.

- A **dependent** variable is the variable that we measure.

- A **continuous** variable, e.g. time or mass, can have any numerical value.

- An **ordered** variable, e.g. small, medium, large, can be listed in order.

- A **discrete** variable can have any value that is a whole number, e.g. 1, 2.

- A **categoric** variable is a variable that can be labelled, e.g. red, blue.

- We use **line graphs** to present data where the independent variable and the dependent variable are both continuous. A line of best fit can be used to show the relationship between variables.

- **Bar graphs** are used to present data when the independent variable is categoric and the dependent variable is continuous.

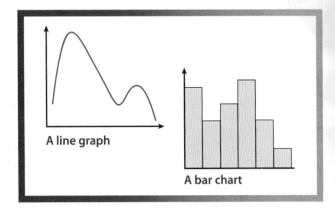

A line graph

A bar chart

Evidence

Evidence should be:

- **reliable** (if you do it again you get the same result)

- **accurate** (close to the true value).

Scientists often try to find links between variables.

Links can be:

- causal; a change in one variable produces a change in the other variable

- a chance occurrence

- due to an association, where both of the observed variables are linked by a third variable.

We can use our existing models and ideas to suggest why something happens. This is called a **hypothesis**. We can use this hypothesis to make a **prediction** that can be tested. When the data is collected, if it does not back up our original models and ideas, we need to check that the data is valid. If it is, we need to go back and change our original models and ideas.

Science in Society

Sometimes scientists investigate subjects that have social consequences, e.g. food safety. When this happens, decisions may be based on a combination of the evidence and other factors such as bias or political considerations.

Although science is helping us to understand more about our world there are still some questions that we cannot answer, eg. is there life on other planets? Or that are for everyone in society, not just scientists, to answer, e.g. should we clone people?

PROGRESS CHECK

1. What is an independent variable?

2. What is a dependent variable?

3. What is an ordered variable?

4. What is a discrete variable?

5. What does accurate mean?

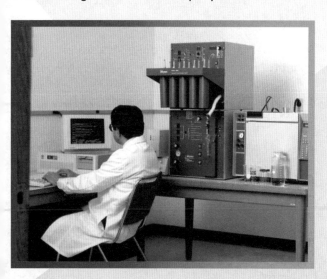

EXAM QUESTION

A student carries out an experiment to find out how the force applied to a spring affects the length of the spring.

a. What is the independent variable?

b. What is the dependent variable?

c. Suggest a variable that must be controlled to make it a fair test.

Motion

The motion of a moving object can be described using measurements of distance, time, speed, velocity and acceleration.

Speed

Speed is a measure of how fast something is going; usually measured in m/s. For example, an object that covers a distance of 3 m every second has a speed of 3 m/s.

$$speed = \frac{distance}{time}$$

- Increasing the speed **increases** the distance covered in the **same** time.

- Increasing the speed **reduces** the amount of time needed to cover the **same** distance.

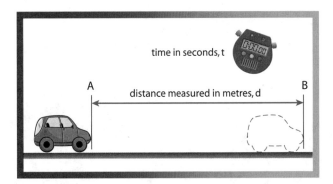

If the car travels a distance AB of 20.0 m in a time t = 5.0 s, its average speed is:

$$\frac{20.0\,m}{5.0\,s} = 4.0\,m/s$$

Speed cameras generally take two photographs a certain time apart and near marked lines on a road. This is because they need to measure the distance travelled in a certain time.

Velocity

Velocity is a measure of how fast something is going in a given direction. Its units are also m/s. The difference between speed and velocity is that velocity includes the direction whereas speed does not.

$$average\ velocity = \frac{displacement}{time}$$

Speed has magnitude (size) only, it is a **scalar** quantity.

Velocity has magnitude and direction, it is a **vector** quantity.

Acceleration

Acceleration is a measure of how fast the velocity is changing. The velocity could be increasing or decreasing.

- If velocity is constant, the change in velocity is zero and the acceleration is 0 m/s per second or 0 m/s/s.

- If velocity increases by 1 m/s every second, the acceleration is 1 m/s/s.

- If velocity increases by 2 m/s every second, the acceleration is 2 m/s/s.

- If velocity decreases by 2 m/s every second, the acceleration is −2 m/s/s; this is often called deceleration.

Acceleration is calculated using the following equation.

$$acceleration = \frac{change\ in\ velocity}{time\ taken}$$

$$or \quad a = \frac{(v - u)}{t}$$

where v = final velocity and u = initial velocity

The unit of acceleration, m/s/s is sometimes written as m/s^2.

To measure acceleration the velocity must be measured at two different times. This can be done with light gates.

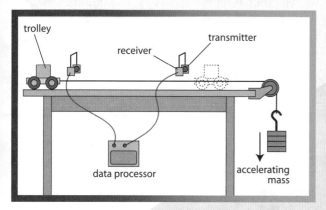

The data processor measures the velocity at two different points and measures the time between them. It calculates the acceleration using the equation: $a = \frac{(v - u)}{t}$.

Sometimes, only the **direction** of a moving body changes. For example, a planet may rotate around the Sun at a constant speed.

In this case:

- speed is constant

- velocity is changing because the direction is changing

- the planet is accelerating because the velocity is changing.

PROGRESS CHECK

1. Why do speed cameras generally take two photographs a certain time apart and near marked lines on a road?

2. What is the equation that relates average velocity, displacement and time?

3. Find the average velocity of a car that travels 300 m in 15 s.

4. Explain the difference between speed and velocity.

5. When a car turns a corner at constant speed, is there a change in velocity?

? EXAM QUESTION

1. State the equation for acceleration.

2. What is the acceleration of a car that accelerates from 10 m/s to 30 m/s in 4.0 s?

3. Find the deceleration of the car if it slows from 30 m/s to 0 m/s in 3.0 s.

4. If a data processor is used to measure the acceleration of a vehicle, what **three** pieces of information does it need?

Calculating Motion

For two cars travelling on a straight road, their relative speed depends on the direction they are travelling.

These two cars are travelling in the same direction at 10 m/s and 12 m/s. Their relative speed is:
10 m/s – 12 m/s = 2 m/s.

These two cars are travelling in opposite directions at 10 m/s and 12 m/s. Their relative speed is:
10 m/s + 12 m/s = 22 m/s.

Vector quantities such as forces or velocities can be added.

These vectors at right angles can be added using the Pythagoras rule for right-angled triangles.

$$\sqrt{(10^2 + 5^2)} = \sqrt{125} = 11\,\text{m/s}$$

Equations

When considering an object whose velocity is changing, values for its initial velocity (u), its final velocity (v) or its average velocity can be measured or calculated. Velocity is always measured in m/s.

There are five equations to calculate motion. Acceleration is represented by a, its units are m/s/s or m/s^2, time (seconds) by t and distance (metres) by s.

$v = u + at$

$\text{average velocity} = \dfrac{(v + u)}{2}$

$v^2 = u^2 + 2as$

$s = ut + \dfrac{1}{2}at^2$

$s = \dfrac{(u + v)}{2} \times t$

Example
A car accelerates from rest at 3.0 m/s/s for 6.0 s. Find the distance it travels.

Use the equation $s = ut + \dfrac{1}{2}at^2$

$u = 0\,\text{m/s}$, $t = 6.0\,\text{s}$ and $a = 3.0\,\text{m/s}$

$s = (0 \times 6.0\,\text{s}) + (\dfrac{1}{2} \times 3.0\,\text{m/s/s} \times 6.0\,\text{s}^2)$

$= 54.0\,\text{m}$

Projectiles

The path of an object projected horizontally is curved due to the downward pull of gravity. The shape of the motion is described as **parabolic**.

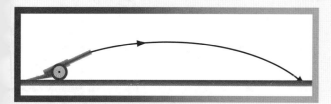

Missiles, golf balls, footballs, netballs and long jumpers are all projectiles. Their path is known as a **trajectory**.

A trajectory's velocity can be separated into a horizontal velocity and a vertical velocity. The resultant velocity is the sum of the vertical and horizontal components.

- ■ The horizontal velocity is constant.

- ■ The vertical velocity steadily increases due to the acceleration of gravity.

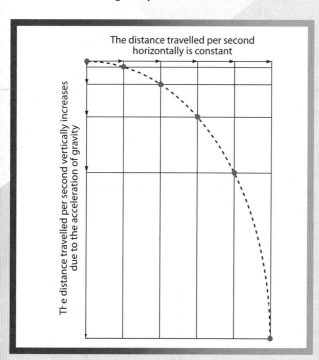

The distance travelled per second horizontally is constant

The distance travelled per second vertically increases due to the acceleration of gravity

Weight is the gravitational force from the Earth, in Newtons. It depends on the gravitational field strength, g, which equals about 10 N/kg on Earth.

Weight = mass × g

An object with mass 20 kg has a weight of 200 N on Earth.

PROGRESS CHECK

1. Find the relative speed of two cars travelling east at 8 m/s and 13 m/s.

2. Find the relative speed of two cars travelling in opposite directions at 8 m/s and 13 m/s.

3. Find the resultant velocity of 5 m/s north and 2 m/s south.

4. Find the magnitude of the resultant velocity of 5 m/s north and 2 m/s west.

5. Find the weight on Earth of 500 g.

EXAM QUESTION

A ball is thrown horizontally from the top of a cliff at a velocity of 2 m/s. It takes 3.5 s to reach the ground.

a. What is the initial vertical velocity?

b. Find the maximum vertical velocity of the ball ($v = u + at$, $g = 10$ m/s/s).

c. Find the height of the cliff ($s = ut + \frac{1}{2}at^2$).

d. How far does the ball travel horizontally in 3.5 s?

Graphs for Motion

Motion graphs show how far something has travelled, how fast it is travelling and how fast it has been travelling along its journey.

Distance–Time Graphs

Distance–time graphs measure distance from a certain point. For example, the distance a car is from a tree.

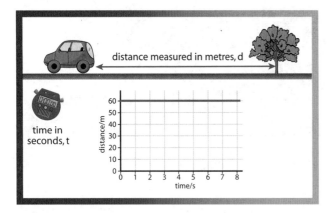

The car is stationary, 60 m away from the tree. There is no change in distance from the tree.

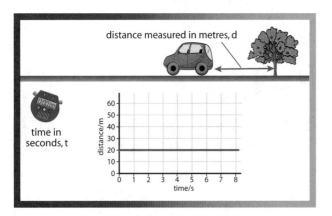

The car is stationary again, this time 20 m away from the tree.

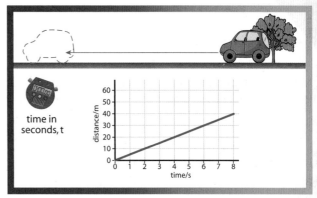

Here, the car is moving. After 4.0 s the car has moved 20 m. After 8.0 s the car has moved 40 m.

The speed of the car is equal to the gradient of the graph. In this case the gradient is $\frac{40\,m}{8.0\,s} = 5.0\,m/s$.

This graph has a steeper gradient, so the speed is greater.

The speed is $\frac{60\,m}{6.0\,s}$ = 10 m/s.

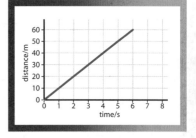

This graph has an increasing gradient, which indicates an increasing speed. The car is accelerating.

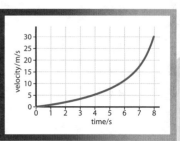

Velocity-Time Graphs

A velocity–time graph shows how the speed or velocity of an object changes with time.

- The **gradient** is equal to the **acceleration**.

- The **area** under the graph is equal to the **distance travelled**.

The velocity–time graph on the right is for the car. It shows a steady speed of 5.0 m/s.

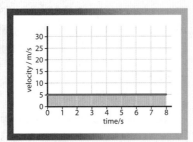

- The gradient is zero so the acceleration is 0 m/s/s.

- The area under the graph is 5.0 m/s × 8.0 s = 40 m. The car has travelled a distance of 40 m.

This graph shows more complicated motion.

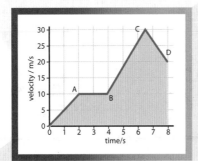

From the origin to A:

- the acceleration = $\dfrac{10\,\text{m/s}}{2.0\,\text{s}}$ = 5.0 m

- the distance travelled = $\dfrac{(10\,\text{m/s} \times 2.0\,\text{s})}{2}$ = 10 m

From B to C:

- the acceleration is $\dfrac{20\,\text{m/s}}{2.5\,\text{s}}$ = 8 m/s/s

- the distance is (10 m × 2.5 s) + ($\dfrac{(20\,\text{m/s} \times 2.5\,\text{s})}{2}$) = 50 m

From C to D the gradient is negative, so the acceleration is negative. The car is slowing down.

1. What does the gradient represent on a distance–time graph?

2. What would a steeper line represent on a distance–time graph?

3. What does the gradient represent on a velocity–time graph?

4. How can you find the distance from a velocity–time graph?

5. What does a negative gradient represent on a velocity–time graph?

? EXAM QUESTION

The graph below shows a skydiver falling through the sky.

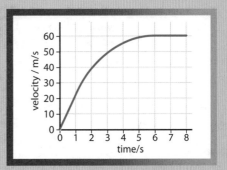

Show on the graph the following points:
a. minimum velocity
b. maximum velocity
c. minimum acceleration
d. maximum acceleration.

When Forces Combine

Forces are vector quantities, they have both magnitude and direction.

Free Body Diagrams

This **free body diagram** shows the forces acting on a girl standing on a plank. There are other forces acting on the plank and on the ground, but the free body diagram only considers the forces that act on a single body (in this case, the girl).

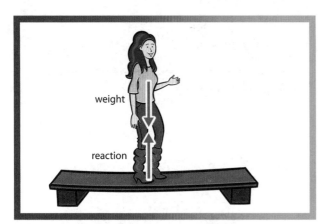

The girl experiences the force of gravity acting down and the reaction force of the plank acting up. These forces are the same size, so they cancel out. There is no net force and the girl stays still. The girl is in equilibrium.

The picture below shows a free body diagram for a plane at constant velocity.

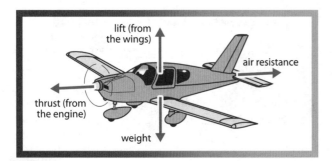

The upward and downward forces on the plane cancel out. The forwards and backwards forces also cancel out. Because the plane is already moving, this means that the plane continues to move at the **same velocity**. The plane is also in equilibrium.

The picture below shows a free body diagram for a parachutist.

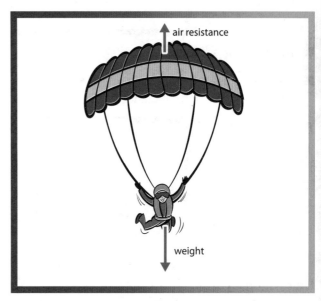

The parachutist experiences weight and air resistance. Since these forces are not equal at the point illustrated, there is a net resultant force downwards.

If weight equals 400 N and the air resistance equals 80 N, the net or resultant force is 320 N in a downward direction. The parachutist is accelerating downwards, he is not in equilibrium.

Note that these two forces are in opposite directions. Therefore the resultant force is found by subtraction. If the forces are in the same direction, the resultant force is found by adding.

Action and Reaction

Whenever two bodies interact, the forces they exert on each other are equal and opposite. An action force produces an equal and opposite reaction force. These forces are always:

- equal in magnitude
- opposite in direction
- the same type of force
- on **different** bodies.

The plank experiences the contact force of the girl and it pushes back with an upward contact force of the same magnitude. Compare the forces in this diagram with the girl on page 12.

Compare the forces in this diagram with the girl on page 12.

PROGRESS CHECK

1. What is a free body diagram?

2. Sketch a free body diagram for a parachutist accelerating.

3. If the weight in your diagram is equal to 550 N and the air resistance is equal to 150 N, what is the resultant force on the parachutist?

4. Consider the man below. Why does the boat move backwards as he steps forwards?

5. Identify the action and the reaction forces for the man and the boat.

EXAM QUESTION

Consider the diagram below.

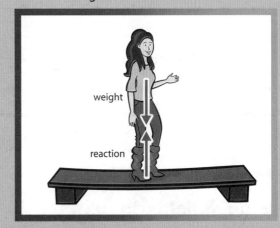

weight

reaction

a. The two forces acting on the girl are equal and opposite. Are they an example of an action and a reaction?

b. Give a reason for your answer.

c. What is the action force that causes the reaction force of the bench upwards on the girl?

d. What is the name given to the above diagram?

Forces and Motion

A force can cause an object to move, it can also cause a moving object to stop moving or change its speed or its direction.

Resultant Force

If the resultant force on a body is zero, it will remain stationary or continue to move at the same speed in the same direction.

If the resultant force on a body is not zero, it will accelerate in the direction of the resultant force.

Acceleration is always caused by a resultant force.

The acceleration can be calculated as follows:

force = mass × acceleration   F = ma

For a given mass:

- more force = more acceleration
- less force = less acceleration

For a given force:

- more mass = less acceleration
- less mass = more acceleration

Terminal Velocity

A falling skydiver experiences a force called weight (gravity) towards the centre of the Earth.

weight = mass × gravitational field strength (g)

On Earth, g is about 10 N/kg.

Initially, weight is the only force. As the skydiver falls, drag, or air resistance, begins to work against the motion.

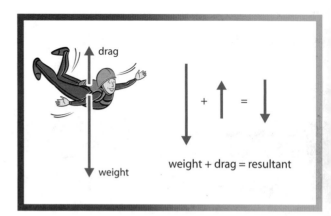

weight + drag = resultant

These two forces combine to produce a resultant force in a downward direction. The skydiver accelerates.

The acceleration produces increased velocity. This produces more drag.

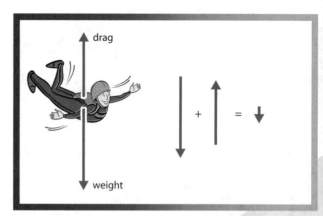

The resultant force is less so the acceleration is less.

The acceleration continues to increase velocity and hence drag. Eventually, drag becomes equal in magnitude to weight.

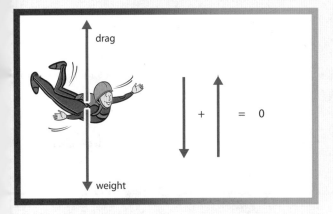

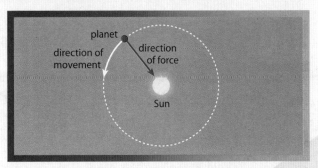

Now the skydiver has reached **terminal velocity**, the maximum velocity attainable. The skydiver continues to fall with constant velocity, there is no further acceleration. The forces are balanced, the skydiver is in equilibrium.

An object with greater surface area, or a less **streamlined** shape has greater drag. For example, a parachutist experiences greater air resistance for the same velocity, he therefore reaches terminal velocity sooner. Terminal velocity will be lower than for the skydiver.

When there is no atmosphere, such as in space, there is no air resistance.

Forces for Circular Motion

An object moving in a circle is accelerating due to its direction changing.

The resultant force that causes this acceleration acts towards the centre of the circle. For a mass on a string, this force is provided by the tension in the string. For an orbiting planet, this force is provided by the gravitational pull on the planet.

The centripetal force needed to make a body perform circular motion increases as:

- the mass of the body increases
- the speed of the body increases
- the radius of the circle decreases.

PROGRESS CHECK

1. A 20 N force is applied to an object of 2.0 kg, what is the acceleration?

2. How would a top box on a car affect air resistance?

3. Does a skydiver or a parachutist experience greater air resistance?

4. What is the direction of the force needed for circular motion?

5. Name **three** ways of increasing the centripetal force required to make a body perform circular motion.

EXAM QUESTION

a. Sketch a diagram of a parachutist accelerating through the air.

b. Label air resistance and weight in your diagram.

c. Under what circumstances does the skydiver stop accelerating?

d. What is the velocity known as at this point?

Momentum and Stopping

The force required to stop a moving object depends on the object's momentum.

Momentum

A faster-moving body has more kinetic energy, it also has more **momentum**.

When calculating what happens to bodies as a result of explosions or collisions, it is often more useful to think in terms of momentum than energy.

> momentum (kgm/s) = mass (kg) × velocity (m/s)

A car of mass 1000 kg travelling at 6 m/s has momentum of 6000 kgm/s.

Momentum has **magnitude** and **direction**.

■ When a force acts on a body that is moving, or able to move, a change in momentum occurs.

■ Momentum is conserved in any collision/explosion providing no external forces act on the bodies.

Consider a trolley of 1.0 kg travelling at 3.0 m/s colliding with a stationary trolley of mass 2.0 kg.

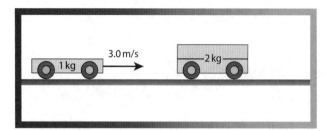

If, after the collision, the first trolley stops and the second trolley begins to move, the velocity of the second trolley can be found.

> momentum before = momentum afterwards
>
> (1.0 kg × 3.0 m/s) + (2.0 kg × 0 m/s) =
> (1.0 kg × 0 m/s) + (2.0 kg × speed after)

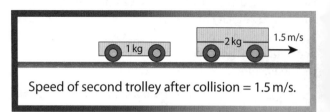

Speed of second trolley after collision = 1.5 m/s.

Momentum is conserved when a rocket is propelled through space. The momentum of fuel ejected in a certain time is equal to the change in momentum of the rocket during that time. Momentum is also conserved in collisions in sport, for example when a racket hits a tennis ball.

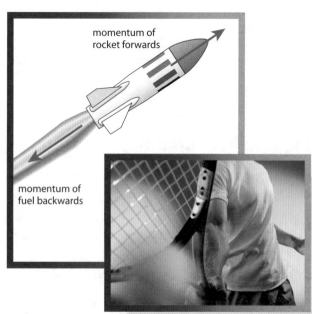

momentum of rocket forwards

momentum of fuel backwards

Stopping

- A vehicle at greater speed requires more braking force to stop in a certain distance.

- A vehicle of greater mass requires more braking force to stop in a certain distance.

The force needed to stop a car is found by considering the change in momentum:

change in momentum = force × time

For a car of mass 1200 kg travelling at 20 m/s stopping, the momentum change is 1200 kg × 20 m/s = 24 000 kgm/s.

To stop in 6.0 s the brakes must exert a force of:

$$\frac{24\,000 \text{ kgm/s}}{6.0 \text{ s}} = 4000 \text{ N}$$

If the braking time is reduced to 3.0 s, the force is doubled to 8000 N.

If the car is stopped more quickly, for example by hitting a tree, the stopping time is reduced and the force is further increased.

Car manufacturers use technology to increase stopping time (and distance) to reduce force on the passengers. Some of the features used are:

- air bags, which inflate, stopping passengers more gently

- seat belts, which stop passengers hitting hard surfaces inside cars

- crumple zones, which collapse steadily in a collision, spreading stopping over a longer time.

When stopping time (and distance) is increased, deceleration is reduced. Injuries are reduced by devices changing shape and absorbing energy.

PROGRESS CHECK

1. Find the momentum of a bicycle and rider of 75 kg travelling at 10 m/s.

2. How does force affect momentum?

3. What is a crumple zone?

4. Define stopping time.

5. How does this affect the passengers?

EXAM QUESTION

a. A trolley of mass 2.0 kg moves with a velocity of 3.0 m/s. Find its momentum.

b. The trolley collides with a stationary trolley of 1.0 kg. The trolleys stick together. Find their combined mass.

c. What is the total momentum of the two trolleys?

d. What is the velocity of the trolleys?

Safe Driving

Driving a car can be a risky business. Many factors can affect the ability of a car to stop.

Stopping Distance

The total **stopping distance** of a car is made up of the **thinking distance** and the braking distance.

- Thinking distance = distance travelled between the need for braking occurring and the brakes starting to act.

- Braking distance = distance taken to stop once the brakes have been applied.

Stopping distance = thinking distance + braking distance

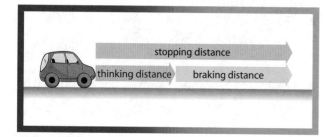

The **thinking distance** depends on the driver's reaction time, which can be affected by the following:

- driver tiredness

- influence of alcohol or other drugs

- distractions or lack of concentration.

A greater speed means that a greater distance will be covered in the same time.

If a car is travelling at a speed of 20 m/s and the driver's reaction time is 0.4 s, the thinking distance is:

distance = speed × time

= 20 m/s × 0.4 s

= 8.0 m

Braking Distance

The **braking distance** can be affected by the following:

- road conditions – friction is affected when roads are wet or icy

- car conditions – bald tyres, braking force

- mass of car

- speed.

Safe drivers consider the above factors and implications on:

- the distance they are from the car in front

- speed limits.

Active and Passive Safety

Active controls make driving safer, prevent crashes from occurring and better protect occupants during a crash.

Some active safety features are:

- ABS braking. To stop quickly, a car needs a high braking force. This relies on the **friction** between the tyres and the road. If brakes are applied too hard the wheels will lock (stop turning) and the car will begin to skid. ABS braking systems detect this locking and automatically adjust the braking force to prevent skidding

- traction control prevents spinning of wheels when excessive acceleration or steering is applied, usually by reducing speed automatically

- safety cages.

Some typical **passive** safety features include:

- electric windows
- cruise control
- paddle shift controls on the steering wheel for changing gears or the stereo
- adjustable seating.

Risk

Risks can be expressed in different ways. People make decisions about the amount of risk they are willing to take. Factors that influence people's willingness to accept risks include:

- degree of familiarity
- whether the risk is imposed or voluntary.

Fuel for Cars

The main fuel in road transport is fossil fuels such as petrol and diesel. Electricity can be used for battery-driven cars and solar cars. These do not pollute at point of use, but need battery recharging that uses electricity supplied from power stations.

Fuel consumption depends on:

- speed
- friction
- driving style
- road conditions.

PROGRESS CHECK

1. Define thinking distance.
2. Define braking distance.
3. State **two** factors that affect thinking distance.
4. What is ABS?
5. Name **three** passive safety features in a car.

EXAM QUESTION

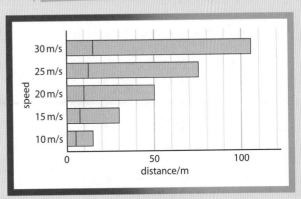

Look at the chart above.

a. At 20 m/s what is the total stopping distance of a car?

b. At 10 m/s what is the braking distance of a car?

c. At 10 m/s what is the thinking distance?

d. State **two** factors that would increase the braking distance.

Energy and Work

Energy and work are both measured in Joules after the physicist James Joule, but they are not quite the same thing.

Transfer of Energy

The scientific meaning of work is the transfer of **energy** from one form to another or from one place to another. A light bulb does work as it changes electrical energy into heat and light. An oven does work as it transfers heat to the food. Therefore, energy is needed to do work.

The law of conservation of energy states that energy can change from one form to another, but it cannot be created or destroyed.

Both energy and work are measured in **Joules** (J).

In this chapter, we will consider **kinetic energy** (KE) and **gravitational potential energy** (PE).

Kinetic energy is found using the equation:

$$\text{kinetic energy} = \frac{1}{2} \times \text{mass} \times \text{velocity}^2$$

$$KE = \frac{1}{2} mv^2$$

All moving objects have kinetic energy. Kinetic energy is greater for objects with:

- greater speed
- greater mass.

Doubling the mass doubles the kinetic energy.

Doubling the speed quadruples the kinetic energy. This means that a car with double the speed needs about four times the braking distance.

Potential energy is found using the equation:

$$\text{potential energy transferred} = \text{mass} \times \text{acceleration of free fall} \times \text{change in height}$$

$$PE = mgh$$

Gravitational potential energy is greater for objects which:

- are more massive
- have more height
- are in a stronger gravitational field.

The acceleration of free fall is sometimes known as gravitational field strength.

Falling Objects

A falling object converts gravitational energy to kinetic energy.

Consider a ball of mass 0.5 kg falling a distance of 15 m.

The potential energy transferred	$= mgh$
	$= 0.5\,kg \times 10\,N/kg \times 15\,m$
	$= 75\,J$

Assuming no energy is lost, all of this energy is converted to kinetic energy.

$$KE = 75\,J = \frac{1}{2}mv^2$$

$$\frac{75\,J}{(\frac{1}{2}\,0.5\,kg)} = v^2$$

$$v^2 = 300$$

$$v - 17\,m/s$$

The speed of the ball after falling 15 m will be 17 m/s.

Weight and Work

Weight is the gravitational force from the Earth. It is measured in Newtons and depends on the gravitational field strength, g, which equals about 10 N/kg on Earth.

$$Weight = mass \times g$$

An object with mass 20 kg has a weight of 200 N on Earth.

PROGRESS CHECK

1. What is the unit of energy?

2. State the equation to calculate kinetic energy.

3. What is the kinetic energy of a mass of 2.0 kg moving at a velocity of 3 m/s?

4. State the equation to calculate weight.

5. What is the weight on Earth of an object of 4.5 kg?

? EXAM QUESTION

Look at the diagram of a rollercoaster ride.

a. What type of energy is gained as the car moves from point A to point B?

b. Calculate the amount of energy gained if the mass of the car is 100 kg.

c. Assuming all of this energy is converted into kinetic energy as the car moves from B to C, how much kinetic energy does the car have at C?

d. Calculate the maximum speed at point C.

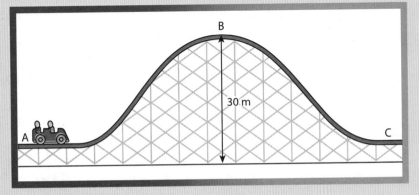

Work and Power

Words such as work and power have special meanings in science and are calculated using specific equations.

Work

Work done is equal to energy transferred; work is done when a force moves an object.

One way of calculating the work done is to use the following equation:

$$\text{work done} = \text{force} \times \text{distance moved (in the direction of the force)}$$

The amount of work done (Joules) depends on:

- the size of the force in Newtons
- the distance moved in metres.

Work is done and power is developed in:

- lifting weights
- climbing stairs
- pulling a sledge
- pushing a shopping trolley.

The man below is running up stairs. He is doing work against his weight, the force of gravity. The distance in the direction of this force is the height of the stairs.

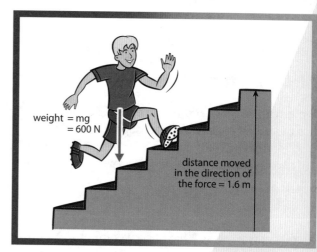

weight = mg = 600 N

distance moved in the direction of the force = 1.6 m

$$\begin{aligned}\text{The work done} &= \text{force} \times \text{distance}\\ &= 600\,\text{N} \times 1.6\,\text{m}\\ &= 960\,\text{J}\end{aligned}$$

He also does some work in the horizontal direction against friction, but this is negligible (small enough to be ignored).

Work done against frictional forces is mainly transformed into heat.

Power is a measurement of how quickly work is being done and is measured in Watts (W).

$$\text{power} = \frac{\text{work done}}{\text{time taken}}$$

f the man above runs up the stairs in 8.0 s his power
s:

$$\frac{960\,J}{8.0\,s} = 120\,W$$

If he walks up more slowly in 20.0 s, his power is:

$$\frac{960}{20.0\,s} = 48.0\,W$$

The faster work is done, the greater the power.

Elastic Potential and Work

For an object that is able to recover its original shape, elastic potential is the energy stored in the object when work is done to change its shape.

PROGRESS CHECK

1. Define work.

2. What is the unit of work?

3. Give **two** examples of when a person is doing work.

4. What work is done when an object is pushed with a force of 50 N for 2.0 m?

5. What is elastic potential energy?

Work Done in Lifting an Object

If an object is lifted a distance of h, the work done is equal to the weight × h:

The weight = mass × g

So the work done in lifting an object: height h = mgh

This is equal to the gain in gravitational potential energy (PE = mgh).

? EXAM QUESTION

Consider the man below running up the stairs. In this question, the force of friction is neglected.

weight = mg
= 500 N

height = 2.0 m

a. What force is he working against?

b. Why is the distance of 2.0 m needed rather than the total distance that he moves?

c. Calculate the work that he does running up the stairs.

d. If he runs faster, does he do more work?

Static Electricity

Static electricity or **charge** is caused by **electrons** – tiny, negative particles around the nucleus of atoms.

When some materials are rubbed together, electrons can be transferred from one material to the other.

■ A build up of electrons creates a negative charge.

■ A lack of electrons creates a positive charge.

You may have experienced static charge when combing your hair. If the comb is made of polythene, it may gain electrons from your hair, leaving your hair positive and the comb negative. Positive and negative charges attract so the comb attracts your hair.

Hair will gain electrons from an acetate comb, leaving the comb positive and your hair negative. Again, the opposite charges will attract.

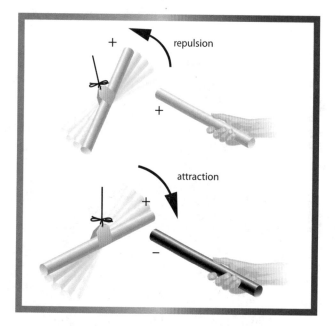

repulsion

+

+

attraction

+

−

Objects that have the same charge will repel.

Insulated materials can be charged by the transfer of electrons. If an object is connected to the ground by a conducting material, the charge flows away.

■ **Conductor** – charge (electrons) can pass through.

■ **Insulator** – charge cannot pass through.

When Static is Dangerous

If enough charge builds up, a spark may jump across the gap between the body and the earth or an earthed conductor.

■ Aircraft fuel causes static as the fuel rubs against the pipe. A spark could ignite the fuel vapour. The aircraft and the tanker are earthed to avoid the charge building up.

■ Lorries with inflammable gases and liquids are earthed before unloading.

■ Lightning happens when charge builds up within a cloud. This can be caused by ice particles rubbing against the air. When the charge is large enough it leaps to another part of the cloud, or to the ground.

When Static is a Nuisance

These situations do not involve enough charge to be dangerous. However, they can be a nuisance.

■ Dirt and dust is attracted to TV screens, monitors and plastic.

■ Synthetic clothing clings.

■ Cars are sometimes earthed to avoid shocks from car doors, caused by friction between you and the car.

■ After walking across a floor covered with insulating material you may experience a shock if you touch water pipes or some other earthed conductor.

Preventing Static Electricity

Some ways to prevent static shocks are:

- correct earthing
- insulating mats
- wearing rubber-soled shoes
- use of anti-static sprays.

Uses of Static Electricity

- Paint spraying – paint is charged so that the droplets repel each other, giving a fine, even spray. A surface is given an opposite charge to the paint, ensuring an even coat and less waste.

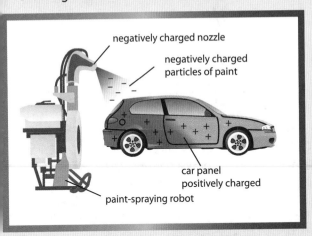

negatively charged nozzle

negatively charged particles of paint

car panel positively charged

paint-spraying robot

- Photocopiers and laser printers use static to direct ink.

- Dust can be removed from chimneys by being attracted to a charged plate. Large particles form, falling when they are heavy enough or shaken.

- Fingerprinting uses static electricity.

- Some dusters are designed to attract dust by static electricity.

- Doctors use static electricity to restart the heart.

Electricity on the Move

When electrons move through a conductor they carry charge. This flow of charge is what we know as electric current.

Electric Current

Electric **current** is the movement of charge through a conductor. Current is measured in **Amps** or Amperes – it is equal to the amount of charge that flows every second. The unit of charge is the Coulomb (C).

This equation relates the amount of electrical charge that flows to current and time.

$$\text{charge} = \text{current} \times \text{time}$$

The charged particles that flow in a wire are electrons, which have a **negative charge**. They flow from negative to positive.

Ions are positive particles left behind when the electrons move. We say that electric current is the flow of **positively** charged ions from positive to negative, but ions don't actually move.

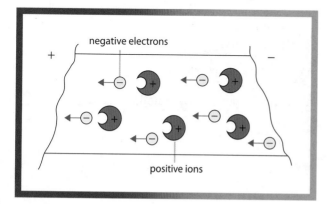

Providing a Potential Difference

Cells produce a flow of charge when a conducting wire is connected across their terminals, caused by the potential difference between the terminals. Potential difference is measured in volts, it is sometimes called voltage. The flow of current from a cell is always in one direction, it is called direct current or DC.

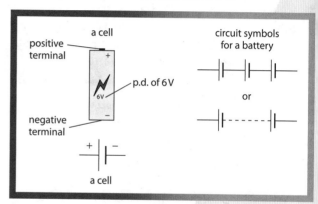

A **battery** is a number of cells connected in series. The number of hours a battery will last depends on its capacity in **Amp-hours** and the current it is supplying. Different appliances will take different currents from the same battery.

$$\text{Hours} = \frac{\text{Amp-hours}}{\text{current supplied}}$$

For example, a battery is labelled 12 Amp-hours and it is used to supply 0.5 A to a CD player.

$$\frac{12 \text{ Amp-hours}}{0.5 \text{ A}} = 24 \text{ hours}$$

The battery will last for 24 hours.

The mains supplies a potential difference of 220 V. The current supplied by the mains is called **alternating current** or **AC** because it continuously changes direction.

Energy

When electrical charge flows through a resistor, electrical **energy** is transformed into heat energy and the resistor heats up.

Energy transformed, potential difference and charge are related by the following equation.

energy transformed = potential difference × charge

In order for a current to flow, a complete loop is required.

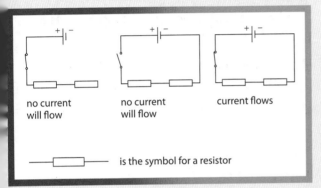

no current will flow

no current will flow

current flows

─▭─ is the symbol for a resistor

Current is measured using an ammeter. The ammeter must be placed in series (next to) any other components. The current is the same at all points in a series circuit so it can be placed either side of the component.

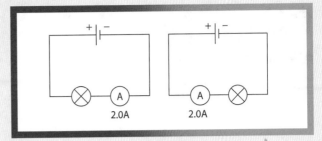

2.0A   2.0A

Measuring Potential Difference

Potential difference is measured using a voltmeter. The voltmeter must be placed in parallel with the component that it is measuring.

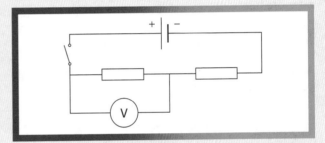

? EXAM QUESTION

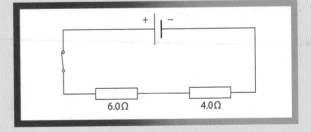

6.0Ω   4.0Ω

a. In the above circuit, electricity flows from positive to negative. Add arrows to the circuit to show the direction of current flow.

b. Add an ammeter to the circuit to measure the current through the resistors.

c. Add a voltmeter that will measure the potential difference across the 4.0 Ω resistor.

d. What will happen if the switch is opened?

◉ PROGRESS CHECK

1. How is current measured?

2. What charge flows per second if the current is 5.0 A?

3. Are the following statements true or false?

   a. Electrons have a negative charge.

   b. Electrons flow from positive to negative.

   c. Ions flow from positive to negative.

   d. Current is said to flow from positive to negative.

Circuits

Resistance is a measure of how easy or how difficult it is for electricity to flow through a material.

Current passes easily through copper wire because it has a low **resistance**. Current does not pass so easily through a filament lamp, however. The filament lamp has a higher **resistance**. More energy is needed to push the electrons through the filament wire in the lamp. This energy is converted to heat (and light) in the lamp. Components with a higher resistance give off more heat.

Calculating Resistance

Resistance is measured in Ohms (Ω). The resistance of a component is found using the following equation:

$$\text{resistance} = \frac{\text{potential difference}}{\text{current}}$$

- Resistance in Ohms (Ω)

- Potential difference (voltage) in Volts (V)

- Current in Amps (A)

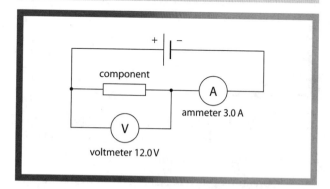

There is a potential difference across this component of 12.0 V and a current through it of 3.0 A.

Resistance $= \dfrac{12.0\,\text{V}}{3.0\,\text{A}} = 4.0\ \Omega.$

- If the potential difference remained the same, a component with a higher resistance would have a *lower* current passing through it.

- If the potential difference remained the same, a component with a lower resistance would have a *higher* current passing through it.

- For a given resistor, current increases as p.d. increases.

- For a given resistor, current decreases as p.d. decreases.

In **series**, the total resistance equals the sum of the individual resistances.

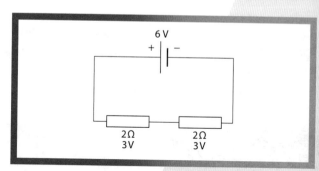

- There is the same current in each component.

- The p.d. of the supply is shared by the components.

n parallel, the total resistance is less than each of the individual resistors.

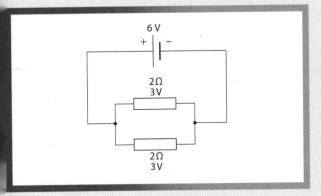

- The p.d. across each component is the same.
- The current is shared by the components.

PROGRESS CHECK

1. Find the resistance of a component with a p.d. of 6.0 V and a current of 4.0 A.

2. A resistor is replaced with a resistor of higher resistance. How will the current in the circuit change?

3. What is the circuit symbol for a variable resistor?

4. What current is drawn by a vacuum cleaner of power 900 W and voltage 220 V?

5. How would decreasing the length of a resistance wire change its resistance?

Changing Resistance

The resistance of some resistors can change. Resistors are usually made from wire where the resistance increases when:

- the wire is longer
- the wire gets hotter.

A **variable resistor** is a device with a control to vary its resistance.

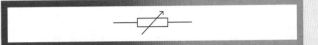

? EXAM QUESTION

In the following circuit, the resistors each have resistance of 5 Ω.

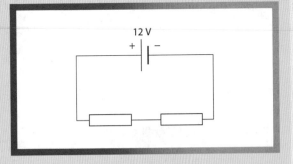

a. What is the total resistance?

b. What is the current through each resistor?

c. What is the p.d. across each resistor?

d. How could both resistors be arranged so that the total resistance is less than 5 Ω?

Electrical Power

Electrical **power** is a measure of the rate at which a device uses **energy**.

An appliance that uses more energy per second has more power. Power is measured in **watts** (W). An appliance that uses 1 joule of energy in 1 second has a power of 1 watt.

A light bulb with a power of 60W uses 60J of energy per second. A light bulb with a power of 100W uses 100J of energy per second.

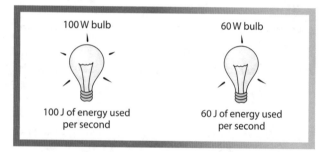

100W bulb   60W bulb

100 J of energy used   60 J of energy used
per second   per second

Power = $\dfrac{\text{Energy}}{\text{time}}$

- Power in watts (W)

- Energy in joules (J)

- Time in seconds (s)

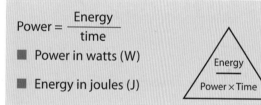

Energy
―――――――
Power × Time

Example

A hairdryer of 1200W (1.2 kW) is used for 2 minutes. How much energy is used?

energy = power × time

= 1200W × 120 s

= 144 000 J = 144 kJ

Calculating Electrical Power

Power is calculated using the following equation:

Power = potential difference (voltage) × current

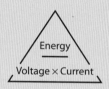

Energy
――――――――――
Voltage × Current

Example

An electric drill operates on a voltage of 230V and has a power of 400W. Calculate the current drawn by the drill.

current = $\dfrac{\text{power}}{\text{voltage}}$

= $\dfrac{400W}{230V}$

= 1.7 A

Domestic Bills

- The standard unit for energy is joules.

- A typical appliance uses a large number of joules, the hairdryer in the example above uses 144 000 J in just 2 minutes!

- For domestic bills, it is more convenient to use a larger unit. The **kilowatt-hour** is a larger unit of **energy**.

- An appliance with a power of 1000 W (1 kW) switched on for 1 hour uses 1 kilowatt-hour of energy. (Compare this to an appliance with a power of 1 W switched on for 1 s which uses 1 J of energy.)

Example

A fridge of 1.5 kW is switched on for 8 hours. How many kWhs does it use?

$$\text{energy} = \text{power} \times \text{time}$$

$$= 1.5\,\text{kW} \times 8\,\text{hours}$$

$$= 12\,\text{kWhs}$$

Electricity companies refer to kWhs as **units** of energy.

The cost can be calculated by multiplying the number of units by the cost of a unit.

cost = number of units (kWhs) × cost of a unit

❓ EXAM QUESTION

A table lamp has three bulbs, each of 40 W.

a. Find the energy in Joules and in kWhs used by each bulb per hour.

b. If each unit costs 10 p, find the cost of lighting the table lamp for 24 hours.

c. It is suggested that the bulbs are replaced with low energy bulbs of 5 W each, giving out the same light, but less heat. How much energy in KWhs is used by each bulb per hour?

d. Calculate the kWhs used by the lamp in 24 hours with low energy bulbs and find the cost of lighting the table lamp for 24 hours.

Electricity at Home

In our homes, many appliances use electricity. When we have a power cut we realise how much we rely on electricity.

Plugs

A plug has three wires: live, neutral and earth.

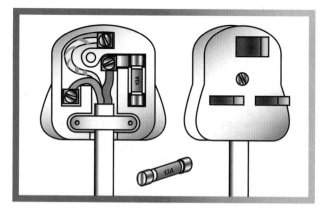

	Colour	Function
Live	brown	alternating between positive and negative voltage
Neutral	blue	completes the circuit, stays at zero voltage
Earth	green/yellow	safety wire

Safe Electricity

Too much current in a wire causes overheating and can cause a fire.

Some ways of improving safety when using electricity are listed here:

- A **fuse** is a short piece of thin wire placed in a circuit that melts with high current, breaking the circuit. The live wire must be connected through the fuse. The power equation can be used to determine the size fuse a device needs. For example, a kettle of 2000 W uses mains voltage of 220 V.

  power = current × voltage

  $$current = \frac{2000\,W}{220\,V} = 9.1\,A$$

  The fuse must be higher, this appliance needs a 10 A fuse.

- **Residual current circuit breakers** (RCCBs) improve the safety of a device by disconnecting a circuit automatically whenever it detects that the flow of current is too high.

- An earth wire is a safety wire that connects the metal parts of a device to earth and stops the device becoming dangerous to touch if there is a loose wire. If a wire works loose and touches a metal part of the device a large current flows to earth, blowing the fuse or activating the circuit breaker.

- Many modern devices are insulated with a plastic case, which insulates the user from any wiring faults. Since the wires themselves are also insulated, this is known as **double insulation**.

Changing Resistance

The resistance of some components is constant. The resistance of other components can change.

- The resistance of a **filament lamp** increases with more current, the lamp gets hotter. The symbol for a lamp is:

filament lamp

A **light dependent resistor** (LDR) has a *high* resistance in the dark and a *low* resistance in the light. It is used to detect light or switch a lamp on automatically in the dark.

light dependent resistor (LDR)

A **thermistor** has a **high** resistance when cold and a **low** resistance when hot. It is used to detect temperature change for a fire alarm.

thermistor

The current through a **diode** flows in one direction only. It has a very high resistance in the reverse direction.

diode

Graphs

If potential difference is plotted on the y-axis and current on the x-axis, the gradient is equal to the resistance. For a component with a constant resistance, for example, a resistor at constant temperature, the gradient is constant.

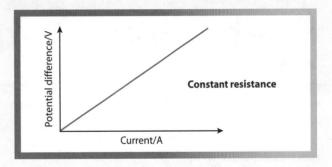
Constant resistance

In a filament lamp, the gradient increases as current increases because the resistance increases as temperature increases.

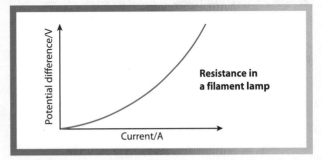
Resistance in a filament lamp

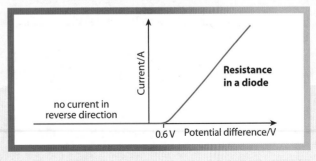
Resistance in a diode

no current in reverse direction

0.6 V   Potential difference/V

PROGRESS CHECK

1. Explain how the resistance of a filament lamp changes with current.

2. What does the graph look like for a component with constant resistance?

3. A thermistor is placed in a beaker of crushed ice. What happens to its resistance?

4. Suggest a use for a thermistor.

5. Suggest a suitable fuse for a vacuum cleaner of power 900W and voltage 220V.

EXAM QUESTION

1. In a plug, what must the live wire be connected through?

2. How is someone using a hairdryer protected by the use of an earth wire and a fuse?

3. What is an RCCB?

4. A hairdryer that is double insulated does not require an earth wire, why not?

Controlling Voltage

Resistors can be used in a circuit to control the voltage in a number of ways.

Two resistors arranged in series can be arranged as a **potential divider**, where:

$$V_{out} = \frac{V_{in} \times R_2}{(R_1 + R_2)}$$

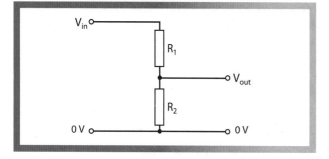

If the input voltage is 12 V and $R_1 = 4\Omega$ and $R_2 = 2\Omega$, the output voltage would be:

$$\frac{12V \times 2\Omega}{(2\Omega + 4\Omega)} = 4V$$

If one of the resistors is replaced with a variable resistor, then the output voltage can be adjusted.

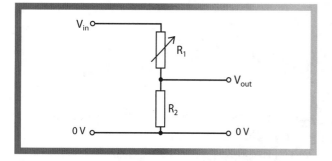

If one of the resistors is replaced with an LDR or a thermistor, the output signal can be used as a sensor of light or heat.

The resistance of a thermistor is high when it is cold, so the voltage across the thermistor is high. In this circuit the output voltage across R_2 is low when cold. When hot, the thermistor's resistance decreases producing a higher output voltage.

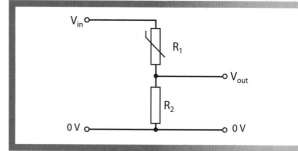

If R_2 is replaced with a thermistor instead of R_1, then the output is reversed.

In the following circuit R_2 has been replaced with an LDR. When it is dark, the LDR has a high resistance and when light it has a low resistance. The output voltage is high when it's dark and low when it's light.

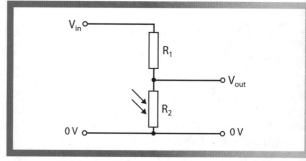

A **diode** can be used to control voltage. A diode only allows electricity to pass through it in one direction. If an alternating current (AC) passes through a diode it becomes a direct current, although it is not smooth like the direct current from a cell.

This is called **half wave rectification**.

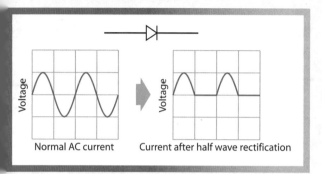

Normal AC current   Current after half wave rectification

A bridge circuit with four diodes produces an AC trace like the one below. This is called full wave rectification.

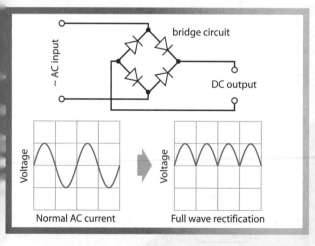

Normal AC current   Full wave rectification

Many devices need a more constant voltage supply. A capacitor can be used to smooth out current that has been rectified in a bridge circuit.

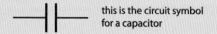

this is the circuit symbol for a capacitor

A capacitor is a device that stores charge. When a current flows in a circuit containing an uncharged capacitor, charge is stored and the p.d. across the capacitor increases.

If a conductor is connected across a charged capacitor, the current flows. As the capacitor discharges, the p.d. across it and the current flowing decreases.

PROGRESS CHECK

1. An AC supply is passed through a single diode. Sketch the shape of the output voltage.

2. Name the process in question 1.

3. Name the process when four diodes are used in a bridge circuit.

4. What is the circuit symbol for a capacitor?

5. Why might a capacitor be included in a bridge circuit?

EXAM QUESTION

1. How does the resistance of an LDR change with light?

2. What is the circuit symbol for an LDR?

3. Sketch a circuit where the voltmeter reading is high in the dark and low in the light.

4. Suggest a use for your circuit.

Electronic Applications

Many electrical devices rely on some form of logic circuit, including computers, washing machines and car ignitions.

The input signal for a logic gate is either a high voltage (about 5V) or a low voltage (about 0V). A high signal is referred to as 1 and a low signal as 0.

NOT gate

The most simple logic gate is a NOT gate which reverses the input. An input of 0 gives an output of 1 and an input of 1 gives an output of 0.

The table below is known as a truth table.

Input	Output
1	0
0	1

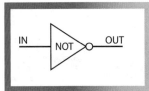

AND gate

An AND gate has two inputs. It gives an output of 1 when both the first and the second input are 1. Otherwise the output is 0.

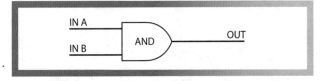

Input A	Input B	Output
0	0	0
0	1	0
1	0	0
1	1	1

OR gate

An OR gate has two inputs. It gives an output of 1 when either the first input or the second input or both inputs are 1.

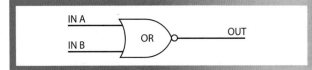

Input A	Input B	Output
0	0	0
0	1	1
1	0	1
1	1	1

An OR gate could be used to control the courtesy light in a car so that it comes on if the driver's door or the passenger's door or both doors are open.

NAND gate

A NAND gate has two inputs and its output is the opposite of an AND gate.

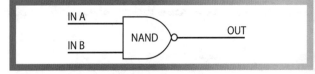

Input A	Input B	Output
0	0	1
0	1	1
1	0	1
1	1	0

NOR gate

A NOR gate has two inputs and is the opposite of an OR gate.

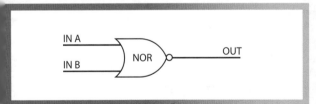

Input A	Input B	Output
0	0	1
0	1	0
1	0	0
1	1	0

A **bistable** arrangement has two stable states. If both inputs are off, then either output C is on (1) and output D is off (0) or vice versa.

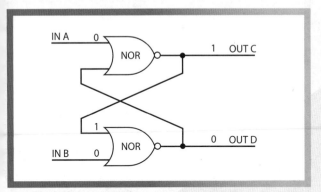

Beginning from the stable state above, if input A is switched on (1) then the bistable flips to the other state, even if input A returns to off (0).

If, when input A has returned to off, input B is switched on, the bistable flips back to the other state. A bistable remembers an input and remains in the corresponding state.

PROGRESS CHECK

1. Draw a NOT gate.

2. Write the truth table for a NOT gate.

3. Draw a NOR gate.

4. Write the truth table for a NOR gate.

5. Suggest a use for an OR gate.

EXAM QUESTION

The logic circuit below is used in an office block.

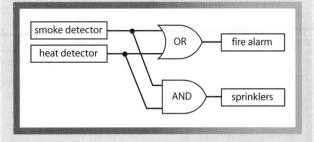

a. What will happen if there is smoke and no heat?

b. What will happen is there is heat and no smoke?.

c. What will happen if there is smoke and heat?

d. Draw a truth table for this system.

Producing Electricity

Moving a wire near a magnet or moving a magnet near a wire generates an electric current in the wire.

Generators

Electricity is generated in a power station by rotating a magnet near a coil of wire (or by rotating a coil of wire near a magnet). This produces a current in the coil of wire and is called the **dynamo effect**.

Dynamos or generators produce a current that changes direction every time the magnet turns. This is called alternating current or AC. The current supplied to our homes is AC.

The current and the voltage of a dynamo can be increased by:

■ increasing the strength of the magnet

■ increasing the number of turns in the coil of wire

■ increasing the speed of rotation of the magnet.

AC and DC current can be displayed on an oscilloscope:

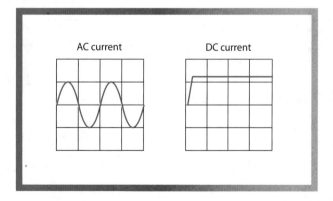

Measuring Current

The current in a circuit is measured using an **ammeter**. The ammeter must be placed in the circuit in **series** (next to) any other components. The current is the same at all points in a series circuit so it can be placed either side of the component.

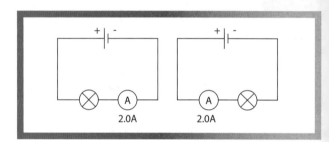

The National Grid

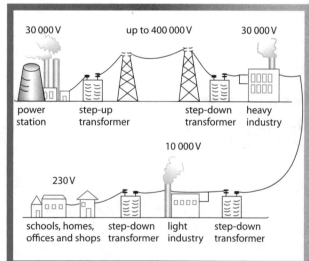

- The National Grid distributes electricity to consumers at different voltages using **step-up** and **step-down** transformers. Consumers include homes, factories, offices, schools and farms.

- It is more efficient to transmit electricity at a much higher voltage than is required by consumers. Increasing the voltage decreases the current in the transmission cables. The cables lose less heat energy when the current is lower.

- Overhead cables are the cheapest way of transmitting electricity. Underground cables are more expensive, but do not spoil the landscape.

- The National Grid allows electricity to be distributed from areas where consumption is low to areas where there is high consumption.

? EXAM QUESTION

The diagram shows a simple generator. When the coil is rotated, a current flows and lights the lamp.

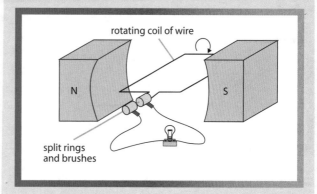

rotating coil of wire

N   S

split rings and brushes

a. Give **three** ways of increasing the current in the coil.

b. Is the current produced AC or DC?

c. Explain the difference between AC and DC.

d. Add to the diagram an ammeter that would measure the current passing through the lamp.

👁 PROGRESS CHECK

1. What is an ammeter?

2. How should an ammeter be connected in a circuit?

3. Why does the National Grid transmit electricity at high voltages?

4. Give an example of a 'high voltage'.

5. Give **one** advantage of using underground cables to distribute electricity.

Motors and Transformers

Electricity is supplied to our homes at 230V, but some devices, such as printers and mobile phone chargers, require less voltage.

Transformers

Transformers change the voltage of an **alternating** electricity supply.

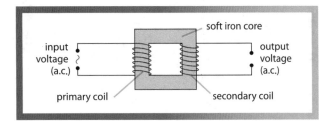

- A **primary** coil is an electric cable wrapped around a soft iron core. It is connected to an input power supply providing **alternating** current.

- The alternating current creates a **changing** magnetic field in the soft iron.

- The changing magnetic field in the soft iron core causes a voltage to be **induced** in the **secondary** or output coil.

- The number of turns of the primary and the secondary coil determines the change in voltage according to the following equation:

$$\frac{\text{primary or input voltage}}{\text{secondary or output voltage}} = \frac{\text{number of turns on primary coil}}{\text{number of turns on secondary coil}}$$

Motors

A current carrying wire has a circular magnetic field around it. The field is made up of concentric circles. When a current carrying straight wire is placed in a magnetic field, it experiences a force and can move.

The direction of this force can be found using Fleming's left-hand rule. A hand is held with two fingers and the thumb perpendicular.

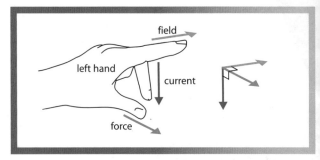

The First finger is pointed in the direction of the Field, the seCond finger in the direCtion of the current and the THumb shows the direction of the THrust or force.

Motors take electrical energy and transform it into kinetic (movement) energy.

This simple motor runs on d.c. (direct current) from a battery. The current flows in the coil, which experiences a turning effect caused by the magnetic field.

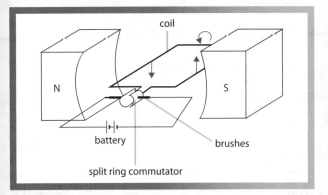

The forces on the coil can only pull the coil half a turn because the forces then pull the wrong way. A motor allows the coil to continue to turn by changing the direction of the forces. It does this by changing the direction of the current, as the coil spins, the commutator spins, but the brushes remain in the same place, changing the direction of the current every half turn.

A stronger turning effect is produced by:

■ increasing the number of turns on the coil

■ increasing the current

■ using a stronger magnet

■ using a bigger coil.

Motors are found in a variety of everyday applications such as washing machines, CD players and electric drills.

PROGRESS CHECK

1. The input voltage of a transformer is 230 V and the primary coil has 1200 turns. If the secondary coil has 100 turns, what is the output voltage?

2. Give **three** ways of increasing the turning effect of a motor.

3. In Fleming's left-hand rule, what does the first finger represent?

4. What does the second finger represent?

5. What does the thumb represent?

EXAM QUESTION

The output of a transformer is 20 V and the input is 200 V.

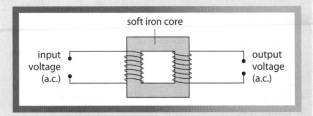

a. What is the ratio of primary turns to secondary turns?

b. Label the primary coil and the secondary coil in the diagram above.

c. Is this a step-up transformer or a step-down transformer?

d. Add a voltmeter to the diagram that would measure the output voltage.

Electricity in the World

People use electricity in many different devices every day without thinking about it.

Everyday Use

Some examples of everyday devices designed to bring about particular energy transformations are listed below.

- A hairdryer converts electrical energy to kinetic energy and heat.
- A microphone converts sound (kinetic energy of the air) to electricity.
- A light bulb converts electricity to light and heat.

The development of the use of electricity and tele-communications has had a great impact on society.

Here are some of the ways that things have changed over time.

- The increased use of electricity and electrical devices has had a dramatic effect on the way we live.
- Electric circuits have become much smaller, greatly increasing the processing speed of computers and allowing development of many new applications.
- The electric telephone has replaced rotary dialling with touchtone dialling and greatly increased the capacity and quality of the network. Digital signalling has replaced analogue, allowing multiplexing (sending many signals at once).
- The use of ICT in collecting and displaying data for analysis has improved reliability and validity of data and encourages faster development of new technology.
- Maglev trains suspend, propel and direct trains using magnetic forces and magnetic levitation from electromagnets attached to the train. The decreased friction allows trains to reach speeds of nearly 600 km/h.

- Superconductivity is the decreasing of resistance to almost zero in certain materials at extremely low temperatures. The technology is used to make very powerful electromagnets used in MRI machines and particle accelerators.

Communication

Mobile phones and satellites use **microwave** and **radio waves**. There is more about satellite communication on page 88.

Lasers produce a narrow intense beam of light in which all the waves are:

- the same frequency
- in phase with each other.

A laser in a CD player reflects from the surface, which contains digital information in a pattern of pits.

Analogue signals have a continuously variable value.

Digital signals are either on or off.

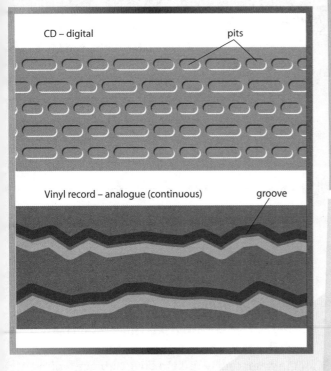

CD – digital   pits

Vinyl record – analogue (continuous)   groove

PROGRESS CHECK

1. How has the size of electric circuits affected computers?

2. What is multiplexing?

3. How can a maglev train reach speeds of almost 600 km/h?

4. What is superconductivity?

5. What is the difference between analogue and digital signals?

EXAM QUESTION

1. Give **two** facts about a laser beam.

2. In what communication device is a laser beam used?

3. Give an example of a digital signal.

Harnessing Energy 1

Fossil fuels are non-renewable, so, eventually, they will run out.

Fossil Fuels

Most of our energy comes from fossil fuels such as oil or gas. Fossil fuels release gases into the environment that pollute the atmosphere.

The energy change when burning fossil fuels is from **chemical energy** to **heat** to **kinetic energy** (in a turbine that turns a generator) to **electrical energy**.

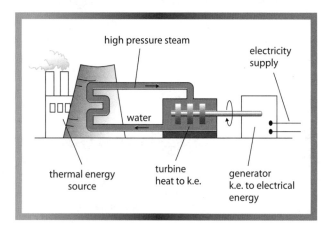

high pressure steam

electricity supply

water

thermal energy source

turbine heat to k.e.

generator k.e. to electrical energy

Energy can be harnessed from sources such as the Sun, the force of the wind or moving water. Such energy sources are renewable, they require no fuel and release no polluting gases into the environment.

There are often difficulties associated with renewable energy sources, however, including other problems for the environment. Many renewable energy sources are also unreliable and expensive. Many people think we should aim to reduce the amount of energy we use in our homes and workplaces.

Wind Power

A wind farm is a collection of generators driven by wind turbines. The generators are used to produce energy.

Some of the problems associated with wind farms are that they:

- require large areas of land
- are dependent on wind speed
- produce noise, which can disturb wildlife
- can be expensive to build
- change the view of the landscape.

The energy change is from **kinetic energy** of the wind to **electrical energy**.

Water Power

Waves and Tides

Energy can be harnessed from the rise and fall of water due to waves and tides. The **kinetic energy** of the water can be used to drive turbines, which generate **electrical energy**.

A disadvantage is that wave and tidal barrages are very difficult to build.

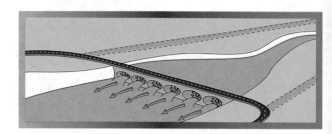

Hydroelectric Power

Rivers and rain fill up a high reservoir behind a dam. The water is released and used to drive turbines that generate electricity. The energy changes are **gravitational potential energy** (of the water) to **kinetic energy** (of the water) to **kinetic energy** (of the turbine) to **electrical energy**.

Harnessing energy from water can disturb natural habitats. Tidal barrages and dams flood areas and remove water from other areas artificially.

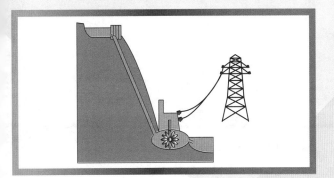

Geothermal Energy

In some volcanic areas, hot water and steam rise to the surface. Water can also be pumped down to hot rocks and steam is generated. The steam can be used to drive turbines. The energy change is from **heat** to **kinetic energy** to **electrical energy**.

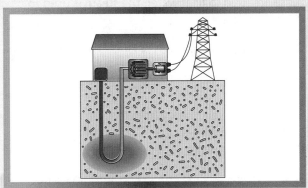

👁 PROGRESS CHECK

1. Name the energy changes in a power station burning fossil fuels.

2. Give **three** renewable energy sources.

3. Name an advantage of hydroelectric power.

4. Name a disadvantage of hydroelectric power.

5. Name the energy change that takes place when using geothermal energy.

❓ EXAM QUESTION

This question is about wind power.

a. Is wind power renewable or non-renewable?

b. Name **two** advantages of wind power.

c. Name **two** disadvantages of wind power.

d. What is the energy change in a wind turbine?

Harnessing Energy 2

Many alternative energy sources disrupt or damage the environment, disturbing wildlife.

Biomass

Fuels can be made from plant or animal matter and these are used to heat water. The steam is used to drive a turbine.

Biofuels require large areas of land to grow the plants.

Solar Power

Solar **cells** convert energy from the Sun into electricity. Solar **panels** use infrared radiation from the sun to heat water.

Solar cells are very expensive and sunshine is not reliable in many locations.

Environmental Issues

- Power stations that burn fuel put carbon dioxide gas into the atmosphere. The gases act like a greenhouse and trap energy from the Sun, which may cause global warming. This greenhouse effect is important – without it our world would not be warm enough for life to exist. Extra warming, however, could cause problems for humans, plants and animals. Deforestation and increased CO_2 also contribute to global warming.

- Some power stations give off other gases such as sulfur dioxide, which can cause acid rain. Acid rain damages buildings and stone.

- A radioactive leak could cause damage to humans and wildlife for many years. Radioactive dust can be carried by the wind for thousands of kilometres.

- The ozone layer protects the Earth from ultraviolet radiation and pollution from CFCs is depleting the ozone layer.

- Dust from volcanoes reflects radiation from the Sun, which causes cooling, dust from factories reflects radiation from the city which causes warming.

Efficiency

$$\text{Efficiency} = \frac{\text{useful energy output}}{\text{total energy input}} \times 100\%$$

All devices waste energy, this energy is usually lost as heat to the surroundings. For example, your body uses energy. If you run up a flight of stairs, you may use 2000 Joules of energy. This energy comes from stored energy in the food you eat. However, only about 15% of this energy is transferred to the movement in your muscles, the rest is wasted as heat in your body and heat to the surroundings. So about 300 J would be transferred to kinetic energy and 1400 J to heat.

Total energy input	= 2000 J
Useful energy output	= 300 J

$$\text{Efficiency} = \frac{300}{2000} \times 100\% = 15\%$$

The sankey diagram/flow chart below shows the energy efficiency for a person running up a flight of stairs.

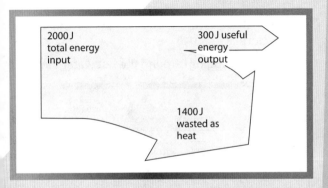

2000 J total energy input

300 J useful energy output

1400 J wasted as heat

Here are some typical efficiencies of other devices.

An electric motor	80%
A petrol engine	25%
A candle	5%
A car engine	25%

A petrol engine loses 75% of its energy as heat, this is why cars need cooling systems.

PROGRESS CHECK

1. State **one** disadvantage of biomass.

2. What chemical causes acid rain?

3. What gas is said to cause the greenhouse effect and global warming?

4. Which gases damage the ozone layer?

5. How does dust from factories affect the climate?

? EXAM QUESTION

a. State the equation to calculate efficiency.

b. If a car burns 4000 J of energy but 3000 J is lost, what is its efficiency?

c. What has happened to the rest of the energy?

d. How do cars prevent overheating?

Keeping Warm

A well insulated house will keep warmer in winter and keep cooler in summer.

Heat travels from hot places to cold places. This generally results in hot things cooling down (losing their heat) and cold things heating up (gaining heat). A warm house in winter can be kept warm using insulation.

Heat can move in three ways:

1. **Conduction** involves the heat moving from one particle to another 'usually in a solid'. Some materials are better thermal conductors than others. Metal is a good conductor since it allows the heat to travel through it quickly. Plastic and wood are poor conductors, or insulators, so are most liquids. Many good thermal insulators have pockets of air trapped in them.

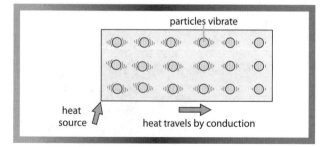

2. **Convection** involves the movement of particles in a gas or a liquid. Thermals are areas of hot air rising, cooler air sinks. The air circulates in a **convection current**. Heat circulates from a heater around a room by convection currents. Convection cannot take place in a solid since the particles are not free to move. Heat transfer by convection can be reduced by trapping air, such as in double glazing.

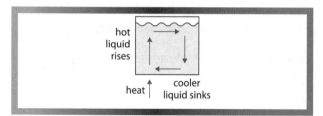

Hot air rises because it expands and becomes less dense (has less mass per unit volume). Cool air sinks because it is more dense.

3. **Radiation** is the transfer of heat energy by **infrared electromagnetic waves**. All bodies emit and absorb thermal radiation. The hotter a body is, the more energy it radiates. Dark, matt objects absorb more radiation than light, shiny ones. This results in dark objects heating up more and emitting more radiation, too. Radiation does not require particles so it can travel through a vacuum. This is how the Sun's energy reaches Earth.

Here are some ways that a house can be insulated. You will notice that many of them trap air and therefore reduce heat transfer by convection.

- Double glazing traps air between two layers of glass.
- Curtains trap air.
- Loft insulation reduces hot air rising and escaping through the roof.
- Cavity walls trap air – cavities in walls are often filled with insulation, which further reduces the movement of the air.
- Draught proofing.
- Reflective foil on or in walls reduces heat loss by radiation.
- Insulation (lagging) around the hot water tank.

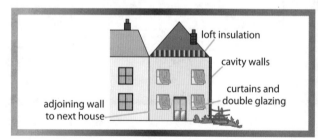

Note that a well insulated house will also keep cooler during a hot summer.

Here are some other factors to take into account when considering the effectiveness of saving energy in a home:

- the thickness of the walls

- the area and number of windows

- a detached house has more surface area than a terraced house.

PROGRESS CHECK

1. Name the **three** ways that heat can move.

2. Which type of heat transfer occurs best in solids?

3. Why does hot air rise?

4. What sort of body absorbs and emits more radiation?

5. Give **three** ways of improving the insulation of a house.

? EXAM QUESTION

The diagram shows a cavity wall which is used to improve the insulation of a house.

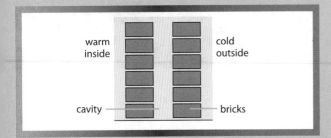

a. By what method does heat flow through the bricks?

b. Draw arrows to show possible convection currents in the cavity.

c. Suggest how the insulation of this system could be improved.

d. Does a cavity wall keep the inside of a house warm or cool when the weather is hot?

Heat

Nearly all of our energy comes from the Sun.

Trapping Heat Energy

Nearly all of our energy comes from the Sun. Solar panels use energy from the Sun to heat up water in pipes to use for heating. **Solar panels** are painted black to increase the absorption of radiation.

Solar cells convert energy from the Sun directly to electricity. **Photocells** convert the Sun's energy directly into electricity.

Some information about photocells is given below.

- Photocells work by absorbing energy that knocks electrons loose from silicon atoms in the crystals.

- Electrons then flow freely to produce direct current.

- The power of photocells depends on their surface area and the light intensity.

- Photocells can operate in remote locations.

- Photocells produce no waste.

- Photocells are low maintenance.

- Photocells do not work in bad weather or at night.

Passive solar heating works because glass allows the Sun's radiation to pass through but it reflects the infrared radiation given off by the heated surfaces inside.

Temperature and Heat Transfer

Temperature is a measurement of hotness and is measured in degrees Celsius (°C). Heat is a measurement of energy and is measured in Joules (J). Temperature can be represented by a range of colours in a thermogram like the two shown here.

Here are some factors that affect the rate at which heat is transferred.

- If a body is warmer (at a higher temperature) than its surroundings, it will lose heat energy to its surroundings. The greater the temperature difference, the faster the rate at which heat is transferred.

- If a body is cooler (at a lower temperature) than its surroundings, it will gain heat energy from the surroundings. The greater the temperature difference, the faster the rate at which heat is transferred.

The shape and the dimensions of a body affect the rate at which it transfers heat. For example, a body with a larger surface area will lose heat faster. Babies have a large surface area compared with their mass; this is why it is more difficult for babies to keep warm.

Under similar conditions, different materials transfer heat at different rates. In a cold room, a carpet does not transfer heat very quickly (it is a good thermal insulator). If you walk on the carpet without shoes you will not feel cold because the heat is being transferred away from your warmer feet at a slow rate. However, If you walk on a tiled floor at the same temperature, your feet will feel cold. This is because the tiles transfer heat away from your feet faster (they are good thermal conductors).

PROGRESS CHECK

1. Why are solar panels painted black?

2. What is temperature?

3. What is heat?

4. Why is it difficult for babies to keep warm?

5. Why does a tiled floor feel cold?

? EXAM QUESTION

This question is about solar cells.

a. What happens to the electrons in a solar cell when it absorbs energy?

b. What factors affect the amount of energy absorbed?

c. Name **one** advantage of solar cells.

d. Name **one** disadvantage of solar cells.

Heat Calculations

The amount of heat gained or lost by a substance can be calculated.

Specific Heat Capacity

The energy needed to change the temperature of a body depends on:

- ▪ its mass
- ▪ the material it is made from
- ▪ the temperature change required.

The **specific heat capacity** tells us how much energy 1 kg of a material needs to increase its temperature by 1 °C. The unit of specific heat capacity is Joules per kg per degree.

The energy required to heat a substance, or the heat lost by a substance when cooling, is calculated using the following equation:

$$\text{Energy} = \text{mass} \times \frac{\text{specific heat}}{\text{capacity}} \times \frac{\text{temperature}}{\text{change}}$$

Example

500 g of water is allowed to cool from 80 °C to 55 °C. How much energy is lost to its surroundings (specific heat capacity of water is 4200 J/kg/°C)?

$$\text{Energy} = 0.5\,\text{kg} \times 4200\,\text{J/kg/°C} \times 25\,\text{°C}$$
$$= 52\,500\,\text{J}$$

When heating a substance, some of the energy is lost to the surroundings. Therefore we calculate the **minimum** amount of energy required.

Specific Latent Heat

If we heat ice at −20°C steadily its temperature increases steadily until it reaches 0°C.

The ice continues to absorb energy at the same rate without getting any hotter, using the energy to break down the bonds between the ice molecules as it melts.

The energy required to melt a substance is the **latent heat**. Liquid requires latent heat to evaporate. Gases lose heat as they condense and liquids lose heat as they freeze. All changes happen at constant temperature.

The **specific latent heat** of a substance is the amount of energy required to change the state of 1 kg of that substance at constant temperature.

We can calculate latent heat using the following equation:

$$\text{Energy} = \text{mass} \times \text{specific latent heat}$$

As a substance is heated or cooled, a graph can show the change in temperature.

The graph on page 53 shows ice as it is heated and melts and then heated (as water) then evaporates.

The heating takes place at a constant rate, so rate of heat added is proportional to the time.

Graph

The horizontal section from 7 to 10 minutes shows that the ice is absorbing heat energy as it changes from ice to water at a constant temperature of 0 °C.

Between 10 minutes and 13 minutes the substance is water.

The horizontal section from 13 minutes to 16 minutes shows the water absorbing heat as it evaporates at 100 °C.

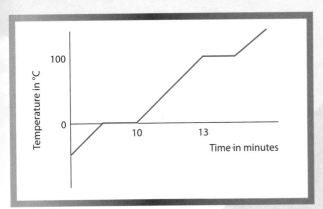

PROGRESS CHECK

1. What is specific heat capacity?

2. At what temperatures does water freeze and melt?

3. What is specific latent heat?

4. Calculate the minimum energy needed to raise the temperature of 2 kg of water by 5 °C.

5. Calculate the energy lost when 1 kg of copper cools by 5 °C (shc = 385 J/kg/°C).

? EXAM QUESTION

Look at the graph above.

a. Explain the horizontal line at 0 °C.

b. Calculate the minimum energy required to heat the water from 0 °C to 100 °C (mass = 1 kg).

c. Why is this the minimum energy required?

d. Calculate the energy required to vaporise the water (latent heat of vaporisation = 2 260 000 J/kg).

Temperature

Temperature is a measurement of hotness and is measured in degrees Celsius (°C).

Heat is a measurement of energy and is measured in Joules (J). Usually temperature is measured in degrees Celsius.

Water freezes and ice melts at 0°C and water evaporates and steam condenses at 100°C.

As particles are heated, the extra energy causes them to move or vibrate faster. As particles lose energy their movement decreases.

At a temperature of −273°C all particles, regardless of their type, stop moving altogether. This temperature is known as **absolute zero**.

The Celsius temperature scale is based on the freezing point and the boiling point of water.

In the 19th Century, a man called Lord Kelvin designed a new temperature scale based on absolute zero. The increments between each point on the Kelvin temperature scale are the same as the increments on the Celsius scale, but zero degrees Kelvin is equal to −273°C, absolute zero.

To change from:

- degrees Kelvin to degrees Celsius, add 273
- degrees Celsius to degrees Kelvin, subtract 273.

The Kelvin temperature of a gas is **directly proportional** to the average kinetic energy of its particles.

The Gas Laws

Pressure in a gas is caused by particles bombarding the surface of the container they are in. Hotter particles move faster and bombard the surface more frequently and with greater force.

For a gas in a sealed container (at constant volume) pressure is proportional to temperature in degrees Kelvin. The unit of pressure is Pascals (Pa).

$$\frac{P}{T} = constant$$

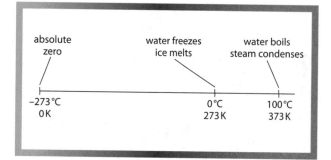

Example

Consider a gas is at a temperature of 15°C at a pressure of 140 000 Pa. If the gas is heated at constant volume to a temperature of 45°C, its new pressure can be found.

15°C = 288 K and 45°C = 318 K

At 288 K $\frac{P}{T} = \frac{140\,000\,Pa}{288\,K} = 486\,Pa/K$

Pressure at 318 K = 486 Pa/K × 318 K = 155 000 Pa

If the gas is not kept at constant volume, the equation below can be used.

$$\frac{P_1 V_1}{T_1} = \frac{P_2 V_2}{T_2}$$

Example

Consider an air bubble underwater at a temperature of 10 °C, a pressure of 200 000 Pa with a volume of 30 cm³.

Find the new volume of the bubble if it rises to the surface where the temperature is 25 °C and the pressure is 100 000 Pa.

T = 25 °C
P = 100 000 Pa

T = 10 °C
P = 200 000 Pa

10°C = 283K
and 25°C = 298K

Since the equation uses ratios, the volume can remain in cm³.

$$\frac{200\,000\,\text{Pa} \times 30\,\text{cm}^3}{283\,\text{K}} = \frac{100\,000\,\text{Pa} \times V}{293\,\text{K}}$$

V = 63 cm³

Remember that these equations only work if the temperature is in degrees Kelvin.

PROGRESS CHECK

1. Define temperature.

2. Define heat.

3. What is the special name given to the temperature −273 °C?

4. What happens to particles at this temperature?

5. What is −273 °C in degrees Kelvin?

EXAM QUESTION

1. How does pressure vary with temperature at fixed volume?

2. Explain this relationship, in terms of particles.

3. State the equation that describes this relationship.

4. If a gas at fixed volume has a pressure of 250 000 Pa at 35 °C, find its pressure at 25 °C.

Inside the Atom

Everything around us including objects, the Earth, the air and even living cells are made of atoms.

An Atom

An atom has a small central **nucleus** composed of **protons** and **neutrons** surrounded by **electrons**.

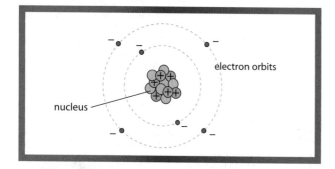

nucleus

electron orbits

Particle	proton	electron	neutron
Charge	+1	−1	none
Mass	1	0	1

The atomic (or proton) number, p, is the number of protons. The mass (or nucleon) number, m, is the number of nucleons (protons + neutrons).

For example, an isotope of Radium (Ra) has 226 nucleons, 88 protons and 138 neutrons.

$$_p^m X \qquad _{88}^{226} Ra$$

- Atoms of the same element have the same number of protons.

- Atoms with neutral charge have the same number of electrons as protons.

- An atom may gain or lose electrons to form charged particles called **ions**.

- Atoms with different nucleon numbers have different mass, although they are still the same element. These atoms are called **isotopes**. An isotope of an element has the same number of protons but extra or fewer neutrons.

Radiation

An unstable nucleus may emit some of its particles. Atoms that give out radiation from their nucleus naturally are said to be **radioactive**.

Alpha radiation is a helium nucleus (2 protons and 2 neutrons). It is emitted straight from the nucleus. For example, radium decays into radon.

$$_{88}^{226} Ra \rightarrow _{86}^{222} Rn + _2^4 \alpha$$

Beta radiation is an electron emitted from the nucleus. No electrons exist as electrons in the nucleus.

However, in an unstable nucleus a neutron may spontaneously change into a proton. When it does this, it emits an electron. This is called beta radiation. For example, iodine decays into Xenon.

$$_{53}^{128} I \rightarrow _{54}^{128} Xe + _{-1}^{0} \beta$$

Gamma radiation is an electromagnetic wave. It usually follows alpha or beta radiation. X-rays have similar properties to gamma rays, but are emitted from different sources.

	Alpha	Beta	Gamma
Range in air	a few centimetres	about a metre	huge distances
Penetration power	stopped by a thick sheet of paper or skin	stopped by a few centimetres of aluminium or other metal	lead or thick concrete will reduce its intensity
Ionising power	strong	weak	very weak
Charge	+2	−1	none

Ionisation is the ability of the radiation to cause other particles to gain or to lose electrons.

EXAM QUESTION

An atom of uranium is written as $^{238}_{92}$U.

a. What is the name given to atoms of the same element with different numbers of neutrons?

b. How many neutrons does each atom of uranium have?

c. The atom emits an alpha particle and decays into thorium (Th), what is the nature of an alpha particle?

d. Write an equation for this decay.

PROGRESS CHECK

1. What is the charge on a neutron?

2. Explain how a beta particle can be an electron emitted from a nucleus.

3. What is the range of gamma radiation in air?

4. What is ionisation?

5. What is the ionisation power of alpha radiation?

Smaller Particles

Neutrons are more difficult to detect than other particles because they have no charge.

If the number of neutrons (N) is plotted against the number of protons (Z) for stable isotopes, the curve below is obtained.

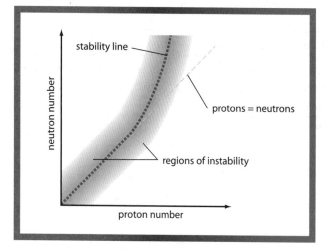

If an isotope does not lie on this curve, it can be identified as unstable and radioactive.

- An isotope above the curve has too many neutrons and will undergo β⁻ decay, emitting an electron.

- An isotope below the curve has too few neutrons and will undergo β⁺ decay, emitting a positron.

- Nuclei that undergo β⁻ or β⁺ decay often undergo rearrangement with a loss of energy emitted as γ radiation.

- Nuclei with more than 82 protons usually undergo α decay.

A **positron** is an electron's anti-particle, identical but with positive charge. In an unstable nucleus a proton may spontaneously change into a neutron plus a positron. The positron is emitted as β⁺ decay.

For example, carbon decays into boron. A particle called a neutrino is also produced.

$$^{11}_{6}C \rightarrow \ ^{11}_{5}Xe + \ ^{0}_{+1}\beta$$

Anti-matter is very similar to ordinary matter, but when matter and anti-matter meet they annihilate each other, releasing energy.

Scientists try to discover more about particles by accelerating particles close to the speed of light and allowing them to collide.

As far as we know, **fundamental particles** are the smallest particles into which matter can be divided.

There are two groups of fundamental particles, quarks and leptons (and their antiparticles, anti-quarks and anti-leptons). An electron is a lepton, a positron is an anti-lepton.

There are six 'flavours' of quarks, up, down, strange, charm, bottom and top.

- Protons can be broken into two up quarks and one down quark, the charges combine to give an overall charge of +1.

- Neutrons can be broken into two down quarks and one up quark, the charges combine to give an overall neutral charge.

- β⁻ decay involves the change of a down quark in a neutron to an up quark. The neutron becomes a proton with a charge of +1 plus an electron.

- β^+ decay involves the change of an up quark in a proton to a down quark. The proton becomes a neutron with no charge plus a positron.

Thermionic emission is when electrons are 'boiled off' hot metal filaments. The following arrangement produces a beam of electrons equivalent to an electric current.

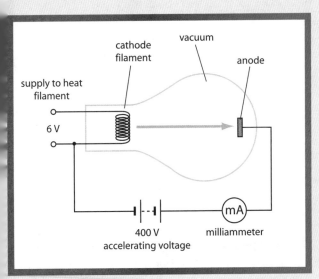

The kinetic energy of the electrons can be found using this equation.

$$KE = \text{electronic charge} \times \text{accelerating voltage}$$

The principal uses of electron beams are:

- TV tubes
- computer monitors
- oscilloscopes
- the production of X-rays.

The electrons in these devices are deflected by an electric field created by parallel charged metal plates. The mass of the particle and the strength of the field affect the deflection.

PROGRESS CHECK

1. What is a positron?

2. Give **two** examples of fundamental particles.

3. What is thermionic emission?

4. State **three** uses of an electron beam.

5. How can electron beams be deflected?

? EXAM QUESTION

a. Sketch a graph of proton number against neutron number.

Show on your graph a possible position of an isotope that:

b. is stable

c. will undergo β^- decay

d. will undergo β^+ decay.

Radioactivity

Radiation is very useful and has improved the quality of human life in areas such as medicine and industry.

Some Uses of Radiation

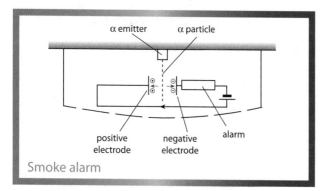

Smoke alarm

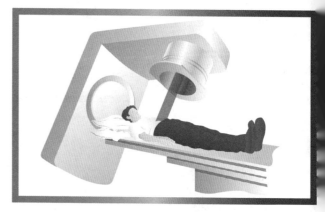

Alpha radiation is used in smoke alarms. Beta or gamma radiation can be used as tracers in industry and for diagnostic purposes in hospitals.

- In hospitals, a tracer is swallowed or injected into the body. It is followed on the outside by a radiation detector. This method is used to check a patient's thyroid gland.

- In industry, tracers are used in underground pipes. A gamma source is normally used so it can penetrate to the surface. The source is tracked by a detector above ground. A leak or a blockage is shown by a reduction in the activity.

- Gamma radiation is used to treat food so it keeps longer.

- Gamma radiation is used to sterilise equipment.

- Gamma rays are used to treat cancer. A wide beam is focused on the tumour and rotated around the person with the tumour at the centre. This limits the damage to non-cancerous tissue. Radiation with the least penetrating effect possible and the shortest possible half-life also minimises exposure.

Half-Life

The radioactivity (or activity) of an isotope is the number of decays emitted per second. It decreases over time.

The half-life of a radioactive sample is defined as the time it takes for the number of undecayed nuclei to halve. It can also be defined as the time it takes for the count rate (activity) to halve. The half-life for any sample is always constant.

A sample with a half-life of 2 days will be half its original value in 2 days, a quarter in 4 days and so on.

The half-life can be used to date artefacts in archaeology and rocks. Archaeologists assume that the amount of Carbon 14 has not changed in the air for thousands of years. When an object (e.g. tree) dies, gaseous exchange with the air stops. The Carbon 14 in the wood decays and the ratio of activity from living matter to the sample leads to a reasonably accurate date. Radioactive dating of rocks depends on the uranium to lead ratio. Scientific conclusions, such as those from radioactive dating, often carry significant uncertainties.

X-Rays

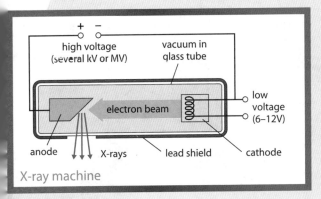

high voltage
(several kV or MV)

vacuum in
glass tube

+   −

electron beam

low
voltage
(6–12V)

anode   X-rays   lead shield   cathode

X-ray machine

A radiographer is a person who takes X-rays. X-rays and gamma rays have similar wavelengths but are produced in different ways.

Gamma rays are given out from the nucleus of certain radioactive materials. X-rays are made by firing high speed electrons at metal targets and are easier to control than gamma rays.

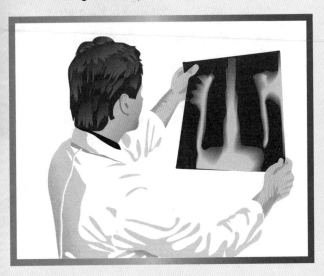

PROGRESS CHECK

1. Suggest a use for alpha radiation.

2. Suggest an industrial use for a tracer.

3. When treating cancer, why is a wide beam of gamma radiation focused on the tumour?

4. What is the difference between X-rays and gamma rays?

5. State an advantage of using X-rays instead of gamma rays.

? EXAM QUESTION

1. Define half-life.

2. A sample has an activity of 400 counts per second. After 3 days the activity is 200 counts per second. What is the half-life?

3. What will the activity be after 9 days?

4. What do archaeologists try to calculate using the half-life of Carbon 14?

Safe Radiation

Radiation can damage or destroy living cells dues to its ionising power. It is important to ensure that people exposed to radiation are kept safe.

Radiation

Radiation can cause cancer or make vital organs stop working.

High levels of radiation pose a greater risk. Over time, scientists have learnt more about the risks associated with radioactive sources.

Background radiation is a low level of radiation that is around us all of the time. It is mainly caused by natural radioactive substances such as rocks, soil, living things and cosmic rays.

Humans also contribute to background radiation, for example medical uses, nuclear waste and power stations.

The Earth's atmosphere and magnetic field protect it from radiation from space. The ozone layer protects the Earth from ultraviolet radiation, but pollution from CFCs is depleting the layer.

Different regions in the UK experience different levels of background radiation. A main cause of this variation is radon gas in the atmosphere.

Nuclear Power

Radioactive fuel rods release energy as heat through a process called **nuclear fission**. A nuclear power station uses this heat to drive turbines and generate electricity.

Nuclear power stations pose two main risks:

- ■ accidental emission of radioactive material

- ■ waste material disposal.

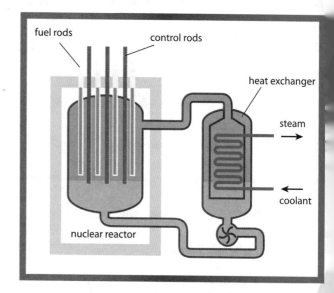

A radioactive leak could cause damage to humans and wildlife for many years. Radioactive dust can be carried by the wind for thousands of kilometres.

Here are some general risks associated with radiation.

- ■ Alpha radiation is highly ionising and very dangerous if it is taken into the body. It cannot pass through the skin. However, if it is absorbed in food or by breathing in radioactive gas or dust, the radiation can cause damage deep inside the body.

- ■ Beta and gamma rays can penetrate the skin but most sources of these radiations are well shielded, such as power stations and laboratories.

- ■ Some media reports claim that microwave radiation from mobile phones or masts poses a health risk.

Safety in the Laboratory

Radioactive substances must be handled safely in the laboratory.

- Tongs or gloves must be used.

- The exposure time must be minimised.

- Sources should be stored in shielded containers.

- Protective clothing must be worn.

? EXAM QUESTION

1. What is the name of the radioactive process that takes place in a nuclear power station?

2. State **two** dangers associated with nuclear power stations.

3. How is the Earth protected from radiation from space?

4. Give **two** precautions that should be taken when handling radioactive sources in a laboratory.

Radiation and Science

Throughout history, scientific ideas and theories about radiation have been changed and developed as new discoveries are made.

Some theories, such as Einstein's theory of relativity, do not originate from experiments. Einstein used his imagination and carried out thought experiments.

Einstein's theory led to predictions which were tested successfully. These tests involved atomic clocks and cosmic rays. The results agreed with the theory and led to more people accepting the ideas.

Without testing new theories, scientists are often reluctant to accept them, especially when they overturn long-established explanations.

Another example of a new theory is 'cold fusion'. Usually fusion requires extremely high temperatures and densities. Theories such as 'cold fusion' are not accepted until they have been validated by the scientific community.

The Plum Pudding Model

Scientists used to believe that atoms were the smallest particles that existed. In 1897, J. J. Thompson discovered tiny, negatively charged particles, which he named electrons. Thompson found that atoms sometimes give out electrons. Since the overall charge of an atom is neutral, Thompson deduced that an atom might consist of a sphere of positively charged mass with negative electrons inside it. This model of the atom is known as the 'plum pudding' model since the electrons look a little like plums in a pudding.

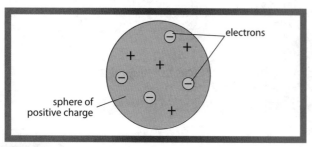

electrons

sphere of positive charge

The Nuclear Model

The scientist Ernest Rutherford had two assistants, Hans Geiger and Ernest Marsden. In 1911, Rutherford asked them to carry out a new experiment to discover more about atoms.

In the experiment, a thin piece of gold foil was bombarded with alpha particles. Gold foil was chosen because it is very thin. Alpha particles are tiny positive particles emitted by some radioactive substances.

Rutherford found that nearly all of the alpha particles went straight through the gold leaf. Very, very few of the alpha particles were deflected by the atoms in the foil!

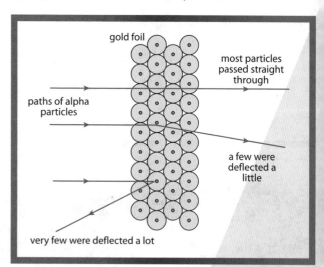

gold foil

most particles passed straight through

paths of alpha particles

a few were deflected a little

very few were deflected a lot

Rutherford concluded that:

■ most of an atom is empty space

■ most of the mass of an atom is compressed into a tiny volume in the centre called a nucleus

■ the nucleus of an atom has an overall positive charge.

Rutherford's new nuclear model of the atom proposed that tiny negatively charged electrons orbited around a dense positive nucleus. The diagram below shows the nucleus many times larger than it would be if the diagram were to scale.

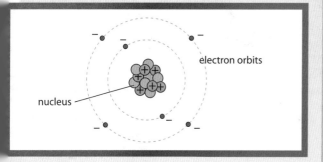

PROGRESS CHECK

1. How does testing a model change the way scientists think about it?

2. Give an example of a recent model that needs testing.

3. What particle did Thompson discover?

4. Why is Thompson's model called the 'plum pudding'?

5. What extra information do we know about electrons today?

Modern Models

Other scientists have refined Rutherford's model with the discovery of neutrons and energy levels for the electrons. Scientists recently discovered that electrons can behave like clouds of charge, or even waves. Today, scientists use a mathematical model of an atom using wave mechanics.

? EXAM QUESTION

This question is about the Rutherford model of the atom.

a. What particles did Rutherford bombard the gold with in his experiment?

b. Why did he use gold foil?

c. What happened to most of the particles?

d. What were the main conclusions of his experiment?

Science and Medicine

Recent developments in the field of physics have improved the quality of care in the field of medicine.

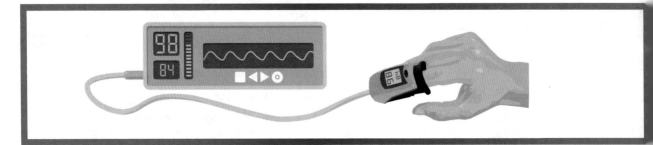

Pulse oximetry is a method used to calculate the amount of oxygen in a patient's blood using LEDs (light emitting diodes) and a photodiode.

The pulse oximeter is connected across a translucent part of a patient's body such as an ear lobe or a fingertip. The LEDs send signals to the photodiode at the frequency of red and infrared light, which is absorbed according to the amount of oxygen in the patient's blood.

Arterial blood vessels expand and contract with each heart beat so the signal varies in time with the heart beat.

Basal Metabolic Rate

Basal metabolic rate is the rate at which energy is used to keep a person's vital organs functioning while at rest.

The following factors affect the basal metabolic rate:
- age
- body mass
- exercise
- nutrition
- temperature
- stress.

Muscle cells can generate potential differences, this can be used in medical applications. For example, the heart generates action potentials as it pumps blood around the body. These action potentials can be measured with an Electrocardiogram (ECG) to monitor heart action.

Below is the characteristic shape of a normal ECG.

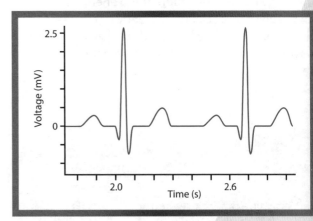

The word **radiation** can be used to describe any form of energy originating from a source. The intensity of radiation will decrease with distance from the source. The intensity can also depend on the nature of the medium through which it is being transmitted.

Intensity can be calculated using the following equation:

$$\text{Intensity} = \frac{\text{power of incident radiation}}{\text{area}}$$

$$I = \frac{P}{A}$$

Radioactive Isotopes

Radioactive isotopes can be created by bombarding certain stable elements with proton radiation to make them into radioactive isotopes. These isotopes usually emit positrons.

When a positron meets with an electron they annihilate one another and produce a gamma ray. This is an example of mass/energy conservation according to Einstein's equation $E=mc^2$. During this annihilation, momentum is conserved.

Positron emission tomography (PET) scanning uses radioactive isotopes with a short half-life, which decay by emitting a positron.

The radioisotope is injected into the body within a chemical. Some time is needed for the chemical to become concentrated in the areas of interest and then the patient is placed in an image scanner.

PET scanning can be used to find cancer tumours in the body. Due to the short half-lives of the radio-isotopes used, they must be created in a cyclotron in the hospital. This makes it very expensive.

Radiation treatment does not always lead to a cure, it is sometimes used to reduce suffering. This is known as palliative care. Ethical issues arise, however, when a patient is sedated extensively to ease suffering – some people believe that this results in a loss of the patient's dignity. Some prefer to forgo treatment for a terminal illness. Some people believe that terminal sedation (euthanasia) is preferable to palliative care or forgoing treatment.

PROGRESS CHECK

1. What does pulse oximetry measure?

2. State **three** things that affect basal metabolic rate.

3. Sketch the normal shape of an ECG.

4. What happens when a positron meets an electron?

5. Why is palliative care sometimes considered unethical?

EXAM QUESTION

1. What do the letters PET stand for?

2. Why do PET scanners use radioisotopes with a short half-life?

3. State a use for a PET scanner.

4. Why is PET scanning expensive?

Nuclear Power

Nuclear power stations produce much more energy than coal- or oil-fired power stations and they emit no greenhouse gases that might contribute to global warming.

Nuclear Fission

Einstein suggested the possibility of releasing enormous amounts of energy trapped in an atom from his relation between mass and energy. **Fission** is the **splitting** of an atom into two lighter nuclei.

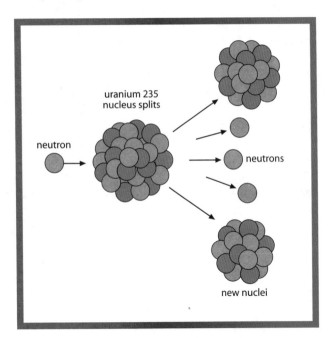

uranium 235 nucleus splits

neutron

neutrons

new nuclei

The process:

- the atom (usually uranium 235) is bombarded with a neutron that is initially absorbed
- this makes the nucleus highly unstable
- the nucleus splits into two lighter nuclei (daughter nuclei) and releases two or three neutrons
- the neutrons bombard other nuclei causing further splitting
- energy is released rapidly.

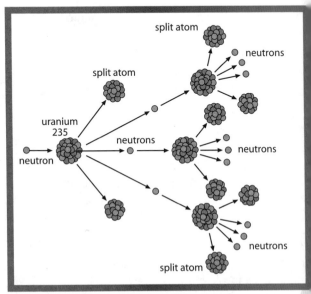

split atom

neutrons

split atom

uranium 235

neutrons

neutron

neutrons

split atom

neutrons

This is known as a **chain reaction**. If the chain reaction is **uncontrolled**, the thermal energy is released extremely rapidly, resulting in an explosion or nuclear bomb.

Nuclear Power

In a nuclear reactor, fission is controlled:

- a moderator such as graphite or water slows down the neutrons
- control rods (that can be raised or lowered) absorb some neutrons but leave enough to keep the reaction going.

A **thermal neutron** has a high kinetic energy and a speed of about 2.2 km/s. If a neutron experiences a number of collisions, it will reach this energy level. Most fission reactors use a moderator to slow down these neutrons so that they interact more easily. Fast breeder reactors use fast neutrons directly.

Radioactive fuel rods such as uranium 235 or plutonium 239 release energy as heat through nuclear fission. In the pressurised reactor below, the hot water is used to produce steam. Steam turns a turbine that drives a generator to produce electricity.

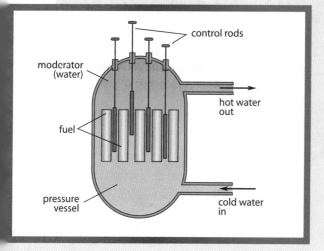

Some of the advantages of nuclear power are:

- it does not contribute to global warming
- it produces high stocks of fuel (non-renewable)
- small amounts of fuel give large amounts of energy (about a million times more than that from burning).

Some disadvantages are:

- it has high maintenance costs
- it has high decommissioning (closing down) costs
- there is a risk of accidental emission of radioactive material
- it creates radioactive waste.

Radioactive Waste

After a few years, the fuel in a reactor must be replaced. The used fuel consists of dangerous radioactive products that need to be stored safely for up to thousands of years. Materials that are near radioactive sources or inside a nuclear reactor absorb extra neutrons, making them radioactive too.

Low-level radioactive waste can be buried in landfill sites. Other waste must be encased in thick glass or concrete and buried. Some types of waste can be reprocessed. Some problems of dealing with radioactive waste are:

- it can cause cancer if not disposed of correctly
- it can remain radioactive for thousands of years
- plutonium (waste product from nuclear reactors) can be used to make bombs and could be considered a terrorist risk
- it must be kept out of groundwater.

Nuclear Fusion

Nuclear fusion is the joining of two atomic nuclei to form a larger one. It is the energy source for stars. It is an impractical source of energy due to the extremely high temperatures and densities it requires. Fission and fusion are forms of neuron radiation.

 PROGRESS CHECK

1. In a chain reaction, a particle is split. What are the products?
2. What does the name 'daughter nuclei' refer to?
3. How can nuclear waste be disposed of?
4. For how long can nuclear waste remain radioactive?
5. What is nuclear fusion?

 EXAM QUESTION

This question is about a nuclear power station.

a. What is a moderator?
b. What are the control rods?
c. State **two** advantages of nuclear power.
d State **two** disadvantages of nuclear power.

Waves

Waves transmit energy from one place to another either through space or through a material.

Transverse and Longitudinal Waves

The wave equation states that for any wave:

$$\text{speed} = \text{frequency} \times \text{wavelength}$$

- **Frequency** is the number of waves passing a point per second and is measured in Hertz (Hz).

- **Wavelength** is the distance between any point on a wave and the same point on the next wave.

- **Amplitude** is the maximum displacement of the waves from rest position.

- **Time period** (T) is the total time taken for a single wave to pass a point. It can be calculated using the equation

$$T = \frac{1}{\text{frequency}}.$$

There are two types of wave, **transverse** and **longitudinal**. Transverse waves vibrate with a displacement **perpendicular** to the direction of travel of the wave. Examples are electromagnetic waves (including light) and water waves.

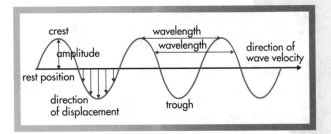

Longitudinal waves vibrate with a displacement **parallel** to the direction of travel of the wave. Examples are sound waves and seismic waves.

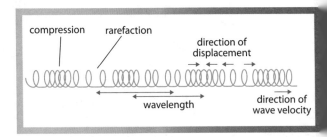

When sound travels through air, the particles vibrate backwards and forwards, parallel to the direction of the wave.

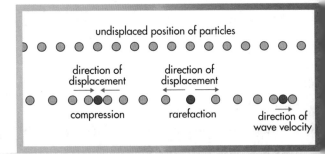

Seismic Waves

Seismic waves are produced by earthquakes as shock waves and can be detected by seismometers.

- P-waves (primary waves) travel through solid and liquid rock and travel faster than s-waves. They are longitudinal.

- S-waves (secondary waves) travel through solid rock only. They are transverse.

Data from seismic waves can be used to draw conclusions about the types of materials that are found inside the Earth. The outer core stops the S-waves because it is liquid. The waves are also refracted within the mantle due to the different densities of the rock.

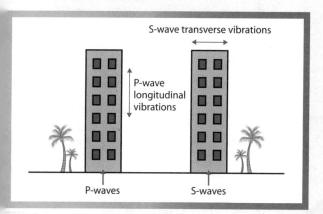

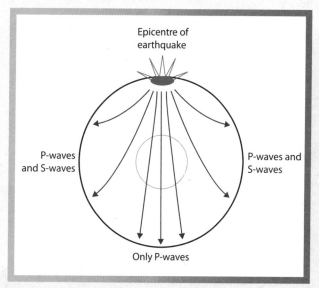

PROGRESS CHECK

1. Define wavelength.

2. Define amplitude.

3. What is the time period of a wave?

4. Which types of waves have compressions and rarefactions?

5. Which types of waves have peaks and troughs?

? EXAM QUESTION

This question is about seismic waves.

a. Name two types of seismic waves produced by an earthquake.

b. Which of these is longitudinal?

c. Which of these can travel through liquid?

d. Explain how these waves can provide information about the Earth's core.

Reflection

We use mirrors every day, they work because of the laws of reflection.

When light is reflected by a plane (straight) mirror, the angle of **incidence** is equal to the angle of **reflection**.

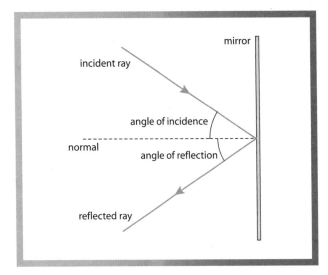

The **normal** is a constructed line perpendicular to any reflecting surface at the point where the incident ray hits the surface.

The nature of an image is defined by:

- whether it is **upright** or **inverted**
- its size relative to the object
- whether it is **real** or **virtual**. A real image can be focused on a screen. A virtual image cannot, you have to look into the mirror to see it.

An image in a plane mirror is:

- upright
- virtual
- the same size as the object.

A curved mirror can be **concave** or **convex**. A concave mirror converges parallel rays so that they all pass through the principal focus, F.

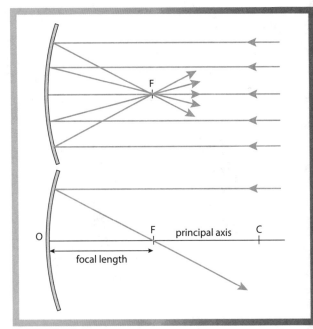

The focal length is half the distance to the radius of curvature C. Both F and C lie on the principle axis. Rays of light reflected at a convex mirror obey the following rules.

- Any ray parallel to the principal axis is reflected through F.
- Any ray passing through F is reflected parallel.
- Any ray passing through C is reflected through C.

The image formed by a concave mirror is found by drawing two rays from a single point on the object and finding where the reflected rays cross. Usually one ray is drawn that passes through C and another that is parallel to the principal axis and reflected through F.

For an object beyond C, the image is:

- real
- inverted
- smaller than the object.

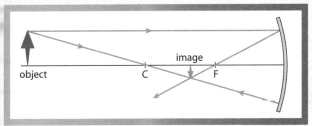

For an object at C, the image is:

- real
- inverted
- same size as the object.

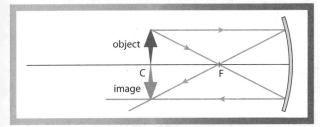

For an object between C and F, the image is:

- real
- inverted
- larger than the object.

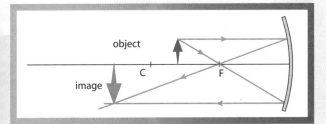

For an object closer than F, the image is:

- virtual
- upright
- magnified.

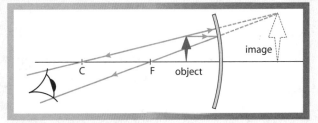

Concave shapes are used for:

- shaving and make-up mirrors
- satellite dishes – microwave signals are brought to a focus
- reflectors in torches and car headlights that form a parallel beam from a light placed at their focal point.

Convex or diverging mirrors form an upright, virtual image that is smaller than the object. Convex mirrors give a wider angle of view and are sometimes used as driving mirrors.

PROGRESS CHECK

1. What law governs reflection in a plane mirror?
2. State **three** properties of an image in a plane mirror.
3. What is a normal?
4. State **two** uses for concave lenses.
5. State a use for a convex mirror.

? EXAM QUESTION

a. Draw a concave mirror with a focal length of 3 cm, mark the radius of curvature on your diagram.

b. Construct an object beyond the radius of curvature.

c. Draw two lines to the mirror to determine the position of the image.

d. State **three** properties of the image.

Wave Behaviour 1

Light travels in straight lines.

There are certain circumstances, however, when light can bend. These include reflection, refraction and diffraction.

Diffraction is the spreading out of waves when they pass through a gap. The waves spread out in all directions with circular wavefronts. Wavefronts are lines drawn one wavelength apart, perpendicular to the direction of the waves. Diffraction does not affect the wavelength.

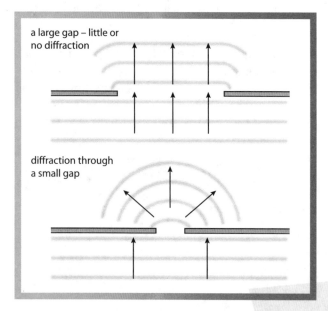

a large gap – little or no diffraction

diffraction through a small gap

Diffraction only happens if the size of the gap is comparable to the wavelength of the wave. The wavelength of visible light is very small, less than 10^{-6}m, this is why visible light does not diffract when it passes through doorways. Long-wave radio waves have wavelengths much larger and will diffract around a hill or over the horizon. Microwave signals to satellites are sent as a thin beam because they will only diffract a small amount due to their short wavelength.

When two waves overlap they produce a pattern of:

- areas where the waves add together (reinforcement)
- areas where the waves subtract from each other (cancellation).

This is known as **interference**. Interference results in:

- louder and quieter areas in sound
- bright and dark areas in light.

A ripple tank produces an interference pattern by making two sets of waves by two dippers vibrating up and down in water powered by the same motor.

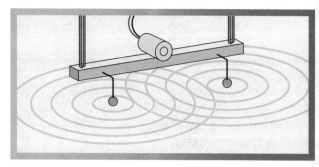

Two sources produce a pattern of constructive and destructive interference.

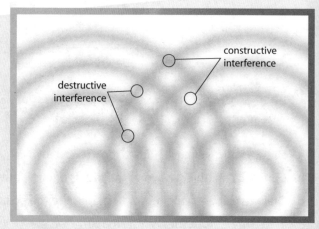

constructive interference

destructive interference

The number of half wavelengths in the path difference for two waves from the same source is:

- an odd number for destructive interference

- an even number for constructive interference.

Polarisation filters out transverse waves leaving only those that oscillate in a single plane.

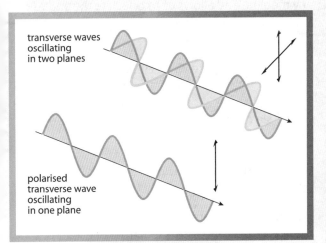

Unpolarised transverse waves such as electromagnetic waves oscillate in all planes, although they can be thought of as oscillating in two planes, horizontally and vertically. After waves have been polarised, all waves have been filtered out except those oscillating in one plane.

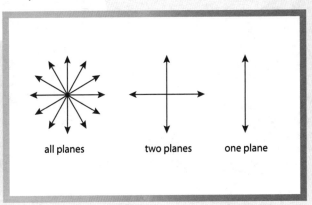

Polaroid sunglasses filter out light in this way. Longitudinal waves cannot be polarised.

PROGRESS CHECK

1. What is a wavefront?

2. Explain constructive interference.

3. For constructive interference, what is the path difference?

4. Can longitudinal waves be polarised?

5. State a use of polarisation.

EXAM QUESTION

1. What is diffraction?

2. What happens to the wavelength during diffraction?

3. Complete the diagram below to show the diffraction patterns for each gap.

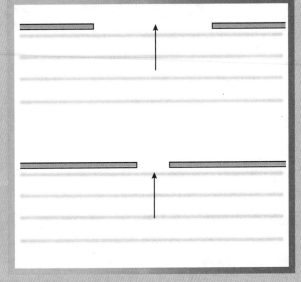

4. Why are the two patterns different?

Wave Behaviour 2

Light travels at different speeds through water, glass and air.

Refraction is when a wave changes direction as it travels from one medium to another. This is a result of a change of speed as the wave enters a new material. For example, water waves slow down when the water depth decreases.

Light slows down when it enters water or glass from air and speeds up when it leaves. Refraction causes swimming pools to look shallower than they really are. This is known as real and apparent depth because the apparent depth is less than the real depth.

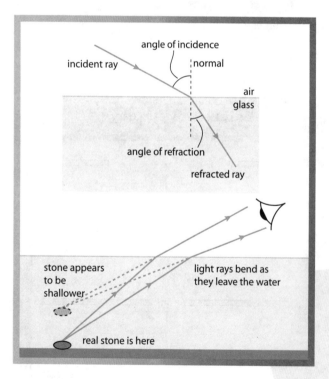

angle of incidence

incident ray — normal

air
glass

angle of refraction

refracted ray

stone appears to be shallower

light rays bend as they leave the water

real stone is here

When a light ray slows down, it bends towards the normal and the angle of refraction is less than the angle of incidence. If it speeds up, it bends away from the normal. If a wave enters a new medium at a right angle, no bending occurs.

The refractive index is known as n and is constant for any medium and allows us to calculate the angles of incidence and refraction.

$$n = \frac{\text{speed of light in vacuum}}{\text{speed of light in medium}}$$

$$n = \frac{\sin i}{\sin r}$$

Where i is the angle of incidence and r is the angle of refraction.

When a narrow beam of white light passes through a prism, it separates into a spectrum of colour due to refraction. White light is made of all of the colours of the rainbow, red, orange, yellow, green, blue and violet. Blue light is deviated more than red. This effect is called **dispersion**.

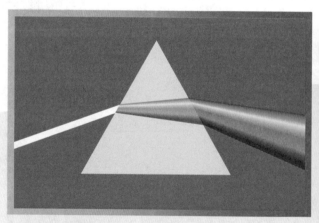

Visible light rays or infrared can be reflected along the inside of an optical fibre if the **incident angle** is greater than or equal to the **critical angle**. This effect is called **total internal reflection**.

The critical angle depends on the refractive index of the medium inside the optical fibre compared to the medium outside (usually air).

$$\sin c = \frac{n_{outside}}{n_{inside}}$$

The higher the refractive index for a material, the lower its critical angle.

Different media have different critical angles. For some types of glass, c is equal to about 41°, for acrylic it is 42°.

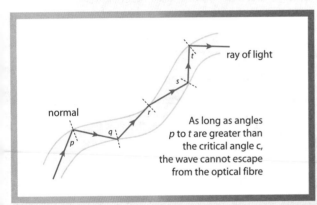

ray of light

normal

As long as angles p to t are greater than the critical angle c, the wave cannot escape from the optical fibre

■ Optical fibres are used in endoscopes – small cameras used to see inside the body.

■ The fibre can be flexible.

■ Optical fibres can transmit data at high speeds.

■ The pulses of light are digital signals.

■ Many digital pulses can be sent on the same data line (multiplexing) allowing more information to be transmitted.

■ Digital signals are clearer, they create less interference.

PROGRESS CHECK

1. What is refraction?

2. Give an example of when light speeds up.

3. Why does a swimming pool seem shallower than it really is?

4. Define dispersion.

5. What is multiplexing?

? EXAM QUESTION

The diagram shows a ray of light entering an optical fibre and reflecting off the inside surface.

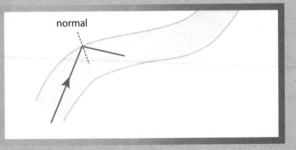

normal

a. What is this effect called?

b. Label the angle that must be greater than the critical angle for this to occur.

c. What is the value of the critical angle for glass?

d. Continue the path of the ray as it reflects along the inside of the glass fibre.

Lenses

Lenses **refract** light that passes through them.

The images they form are similar to those formed by curved mirrors. A real image can be focused on a screen. Cameras and projectors form real images.

A **convex** or converging lens forms different images depending on the position of the object. The position of the image is found by drawing two rays from the image and finding the position where they cross.

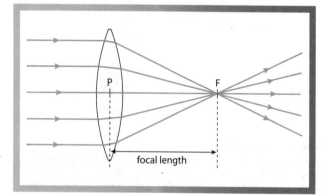

Refracted rays obey the following rules.

- Any ray passing through the **optical centre** of the lens, P, is undeflected.

- Any ray parallel to the principal axis is refracted through the focal point, F.

- Any ray passing through the focal point is refracted parallel to the principal axis.

An image formed by an object beyond 2F is:
- real
- inverted
- smaller than the object.

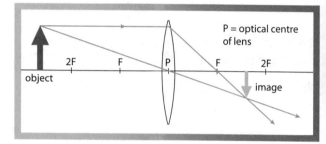

An image formed by an object at 2F is:
- real
- inverted
- same size as the object.

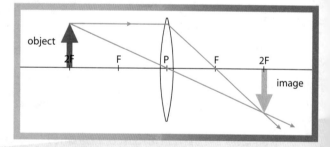

An image formed by an object between F and 2F is:
- real
- inverted
- larger than the object.

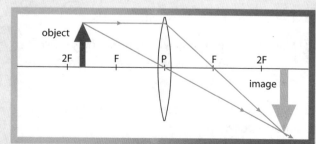

An image formed by an object that is closer than F is:

- virtual
- upright
- magnified.

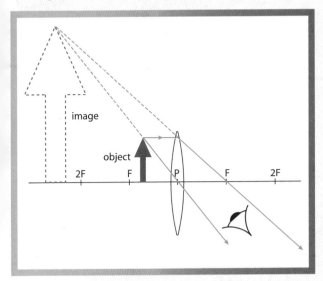

A fatter lens made of the same material as a thinner lens has a shorter focal length. Converging lenses are used:

- to focus an image on a film in a camera
- in a magnifying glass
- in projectors.

Magnification

A ray diagram can be carefully constructed to scale in order to find the magnification of the image. The magnification of an image formed by a curved mirror or a lens can be found using the following equation.

$$\text{Magnification} = \frac{\text{image height}}{\text{object height}}$$

A **concave** or diverging lens forms an upright virtual image of any object. The image is always closer to the lens than the object and smaller.

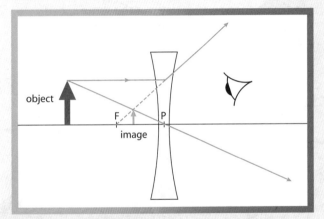

The dashed ray is a virtual ray drawn back to the focal point in order to find the position of the image.

PROGRESS CHECK

1. Give another name for a convex lens.

2. Define the focal length of a lens.

3. State **three** properties of an image formed by a convex lens if the object is placed at 2F.

4. What is the difference between a real image and a virtual image?

5. Give **two** properties of an image formed by a concave lens.

? EXAM QUESTION

a. Sketch a convex lens with a focal length of 2 cm.

b. Place an object 1 cm high, 1.5 cm from the optical centre of the lens.

c. Draw lines to enable you to determine the position and size of the image.

d. Draw an eye to show the position an observer must be in order to view the image.

The Electromagnetic Spectrum

Microwaves, X-rays, gamma rays, infrared and visible light are all electromagnetic waves.

Electromagnetic Radiation

There are many types of **electromagnetic radiation**, including visible light. Different colours of visible light have different frequencies and wavelengths. Other electromagnetic waves have a wider range of frequencies and wavelengths outside the range of detection of human eyes; they are invisible. All electromagnetic waves travel at the same speed through a vacuum (or air), the speed of light $(300 \times 10^6 \, \text{m/s})$.

The Electromagnetic Spectrum

The electromagnetic spectrum is continuous, but waves within it are grouped into types as shown below.

Different wavelengths of electromagnetic waves are reflected, refracted, absorbed or transmitted differently by different substances and types of surfaces.

Excessive exposure to **electromagnetic** waves can be detrimental. Higher-frequency radiation causes greater risk.

- **Infrared waves** are given off by heaters, too much can cause skin burns. Particles at the surface absorb infrared, the heat is transferred to the centre by convection or conduction. Infrared sensors are used in remote controls, burglar alarms and short-distance computer data links.

- **Ultraviolet light** is emitted naturally by the Sun. Overexposure damages skin and eyes and causes cancer. High factor sunblock reduces the risks.

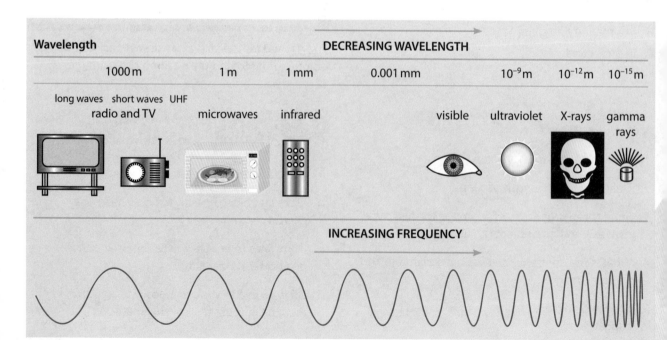

Wavelength			DECREASING WAVELENGTH →				
1000 m	1 m	1 mm	0.001 mm	10^{-9} m	10^{-12} m	10^{-15} m

long waves short waves UHF
radio and TV   microwaves   infrared   visible   ultraviolet   X-rays   gamma rays

INCREASING FREQUENCY →

- **Microwaves** are absorbed by particles on the outside layers of food to a depth of about 1 cm, increasing the kinetic energy of the particles. The heat is transferred to the centre by conduction or convection. Microwaves are reflected by metal, but can go through glass and plastics. Microwaves of high frequency and energy can cause burns when absorbed by body tissue.

- **X-rays** and **gamma rays** cause mutation or destruction of cells in the body.

Absorption and Reflection

Different wavelengths of electromagnetic radiation have different effects on living cells. Some pass through, some are absorbed and produce heat or an alternating current of the same frequency of the radiation. Some cause cancerous changes and some may kill the cells. These effects depend on the type of radiation and the size of the dose.

Scanning by **absorption**:

- microwaves used to monitor rain
- X-rays to see bone fractures
- ultraviolet light to detect forged bank notes.

Scanning by **reflection**:

- ultrasound to scan a foetus during pregnancy. The distance to the reflecting surface is calculated using:

$$speed = \frac{distance}{time}$$

- optical waves for iris recognition.

Radio frequencies below 30 MHz are reflected by the ionosphere. Above 30 GHz, rain, dust and other atmospheric effects reduce the strength of the signal due to absorption and scattering.

Infrared uses scanning by **emission** to monitor temperature.

PROGRESS CHECK

1. Some electromagnetic radiation passes through living cells, some is absorbed, some causes cancerous changes. What **two** properties of the radiation effect what happens?

2. Explain another possible effect of radiation on living cells.

3. State **one** application of scanning by absorption.

4. State **one** application of scanning by reflection.

5. State **one** application of scanning by emission.

EXAM QUESTION

1. Name **four** types of electromagnetic radiation.

2. Suggest a typical wavelength for gamma rays.

3. Suggest a typical wavelength for microwaves.

4. How do microwaves heat food?

Sound

Sounds are mechanical vibrations that can be detected by the human ear.

The human ear can detect mechanical vibrations if they are in the frequency range 20–20 000 Hz.

Sound travels as a wave, usually through air and is a longitudinal wave. Sound waves can be reflected and refracted.

When sound travels through air, the particles vibrate backwards and forwards, parallel to the direction of the wave.

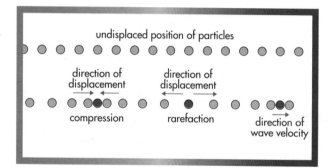

The longitudinal vibrations can be displayed by an oscilloscope using a microphone. The oscilloscope translates the vibrations into an image of a transverse wave. The amplitude, the wavelength and the frequency of the wave can be observed on the trace.

These two waveforms represent sound of different wavelengths and frequencies. The waveform with a high frequency has a shorter wavelength and the waveform with a low frequency has a longer wavelength.

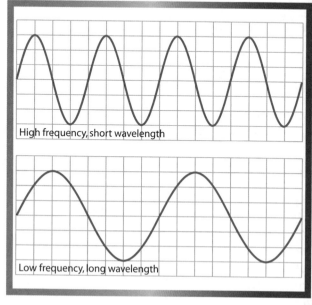

High frequency, short wavelength

Low frequency, long wavelength

The pitch (note) of a sound wave depends on the frequency. A high-frequency sound wave has a high pitch (note) and a low-frequency sound wave has a low pitch.

The two waveforms on the right represent sound waves of identical frequencies and wavelengths but different amplitudes.

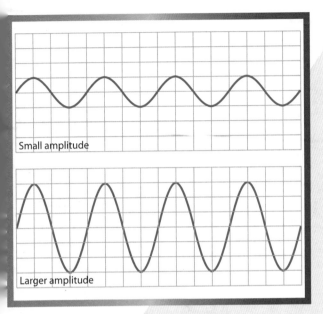

Small amplitude

Larger amplitude

The amplitude of a sound wave determines the volume or loudness of the note. A wave with a small amplitude has a soft or quiet note and a wave with a large amplitude is loud.

Frequency and amplitude are independent. Therefore pitch and volume are independent. This means that a high pitch note can be loud or soft and a loud sound can have a high pitch or a low pitch.

Ultrasound

Ultrasound waves have higher frequency than the upper threshold of human hearing. That means that the frequency of ultrasound is above 20 000 Hz.

Ultrasound waves are partially reflected when they meet a boundary between two different media. The time taken for reflections to reach a detector is a measure of how far away the boundary is.

Ultrasound is used to:

- look inside people (body scans), it reflects at layer boundaries
- pre-natal scanning
- break down kidney and other stones
- measure the speed of blood flow
- clean and control quality in industry.

Ultrasound has advantages over X-rays in that it:

- can produce images of soft tissue
- does not damage living cells.

PROGRESS CHECK

1. Is sound a transverse wave?

2. What name is given to sound waves of frequency above the threshold of human hearing?

3. What effect does the amplitude of a sound wave have on the sound it produces?

4. What effect does the frequency of a sound wave have on the sound it produces?

? EXAM QUESTION

This question is about ultrasound.

a. State a possible frequency for ultrasound.

b. How is the distance to a boundary measured using ultrasound?

c. State **three** uses of ultrasound.

d. State **two** advantages of ultrasound compared to X-rays.

Beyond our Planet

Our star is one of millions of stars in our galaxy, the Milky Way. There are also millions of galaxies in our Universe.

The Universe

The universe consists of:

- stars and planets
- comets and meteorites
- black holes
- large groups of stars called galaxies.

Our **galaxy** is called the Milky Way and consists of billions of stars. In our solar system, the Earth is one of many planets orbiting our star (the Sun) with different radiuses and time periods. Gravity provides the centripetal force for orbital motion.

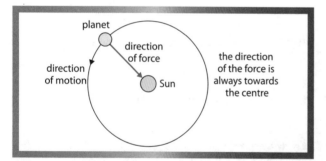

The Earth has an orbit of 365 days. Every 24 hours, it rotates on its axis creating day and night. The existence of life on our planet is determined by the position of our planet within our solar system and the position of our star, the Sun, in its life cycle.

The Moon

The Moon orbits Earth. The Moon may be the remains of a planet that collided with the Earth. When two planets collide, their iron cores can merge to form a larger planet and the less-dense material orbits as a moon.

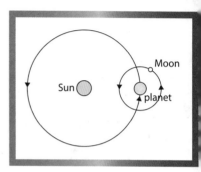

Comets

Comets orbit the Sun with highly elliptical orbits. They are made from ice and dust from far beyond the planets. A comet has least speed when furthest from the Sun, when the gravitational pull is the weakest.

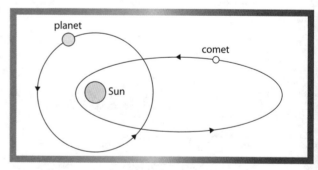

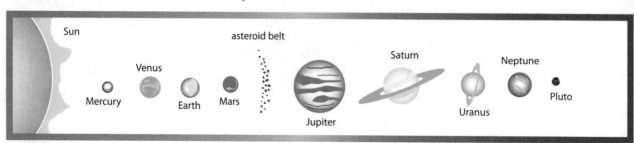

Asteroids

Asteroids are thousands of rocks orbiting the Sun, left over from the formation of the solar system. The large gravity of Jupiter disrupts the formation of a planet, causing an asteroid belt to orbit between Mars and Jupiter.

Consequences of asteroid collision can be:

- craters
- ejection of hot rocks
- widespread fires
- sunlight blocked by dust
- climate change
- species extinction (causing sudden changes of fossil numbers).

A **near-earth-object** (NEO) is an asteroid or a comet on a possible collision course with the Earth. An NEO may be seen with telescopes, monitored by satellites and deflected by explosions.

Stars

Stars are huge balls of burning gas that give off their own light. Here is the life history of a star.

- Interstellar gas cloud.
- Gravitational collapse producing a protostar.
- Thermonuclear fusion.
- A long period of normal life (**main sequence**).

At the end of its life, the outer layer of a medium-weight star expands to become a **red giant**. It then drifts into space as planetary nebula leaving a **white dwarf**.

A heavyweight star also becomes a red giant, then explodes as a **supernova**, then becomes a very dense **neutron star** or a black hole. Our Sun is a medium-weight main sequence star.

A **black hole** has a very large mass and very strong gravity. Nothing can escape from a black hole, not even light. Much of the universe is made up of **dark matter**, we know very little about its nature.

👁 PROGRESS CHECK

1. How often does the Earth rotate?

2. What is the name of our galaxy?

3. What is a red giant?

4. Describe the end of a life cycle of a very massive star.

5. Why can light not escape from a black hole?

❓ EXAM QUESTION

Our solar system consists of nine planets orbiting the Sun.

a. Name **two** other things that orbit the Sun.

b. Why is the asteroid belt between Mars and Jupiter?

c. Give **three** possible consequences of collision with an asteroid.

d. What is a moon?

Space Exploration

Space has no gravity, no atmosphere and varies dramatically in temperature.

Extended periods in space can cause deterioration of bones and heart and can subject an astronaut to dangerous levels of radiation.

A manned spacecraft must provide:

- enough fuel
- shielding from cosmic rays
- heating and cooling
- artificial gravity
- air supply
- enough food and water
- exercise machines.

Space travel takes a long time and covers huge distances. **Light years** are used to measure these distances, one light year is the distance light travels in a year.

An unmanned spacecraft can withstand conditions that humans cannot. They have the advantages of cost and safety. However, if something goes wrong, it is difficult to fix. Unmanned spacecraft can collect information on:

- temperature
- magnetic field
- radiation
- gravity
- atmosphere.

Mass, Weight and Force

The **mass** of a body is how much matter it is made of, it does not change. Mass is measured in kilograms (kg). The **weight** of a body is the amount of gravitational **force** acting on it and can be calculated as follows:

Weight (N) = mass (kg) × acceleration of free fall (N/kg)

The acceleration of free fall (g) is about 9.8 N/kg on Earth.

A spacecraft is powered by ejecting fuel backwards. The **action** of the fuel backwards causes the **reaction** of the spacecraft forwards. Its motion can be predicted using the equation:

Force (N) = mass (kg) × acceleration (m/s/s)

Observations of our universe can be carried out on Earth or in space. Telescopes can detect visible light or other non-visible electromagnetic waves.

Big Bang Theory

One of the theories of the origin of the universe is that it began with a **Big Bang** from a very small initial point. When galaxies move away from us at high speeds, light waves from them become stretched out, moving their wavelength closer to the red end of the visible spectrum, called 'red-shift'. The more distant galaxies have a greater red-shift. This means that the galaxies are moving away from each other and the universe is expanding. This supports the Big Bang theory and predicts that the universe began about 15 000 million years ago.

The faint microwave radiation from space, known as cosmic rays and picked up by telescopes, may be left over from the Big Bang.

It is thought that the universe may eventually stop expanding and reach a **steady state**, or that it might eventually collapse and then expand again, **oscillating** continuously.

👁 PROGRESS CHECK

1. How can space travel affect the human body?

2. State **three** things a manned spacecraft must provide.

3. What is a light year?

4. Give **three** things that an unmanned spacecraft can collect information about.

5. What **two** pieces of evidence are there to support the Big Bang theory?

❓ EXAM QUESTION

1. Explain the difference between mass and weight.

2. What is the weight on Earth of an object of mass 7.5 kg? (g=9.8 N/kg on Earth)

3. The object is taken into space in a rocket: state the effects, if any, on the mass and the weight of the object.

4. The rocket is propelled forwards by the reaction force caused by the rocket's fuel. Which direction is the rocket's fuel projected?

Communication

Every time a mobile phone is used, signals are sent into the upper atmosphere, or to satellites orbiting the Earth.

Communication Signals

Mobile phones and satellites use **microwave** and **radio waves**. Signals can be reflected from the ionosphere or received and re-transmitted by satellites. Diffraction, refraction and interference around obstacles cause signal loss.

This can be reduced by:

- limiting the distance between transmitters
- positioning transmitters as high as possible.

Transmission dishes also experience signal loss due to diffraction.

There have been claims that using mobile phones or living near a phone mast may be dangerous.

Magnetic Fields

The Earth is surrounded by a magnetic field. This is because the Earth's core contains a lot of molten iron. Magnets have a North and a South pole, which can be identified using a plotting compass.

The magnetic field around our planet shields us from much of the ionising radiation from space. An electrical current in a coil also creates a magnetic field.

A current flows in the coil (also called a solenoid) producing a magnetic field. The field is the same shape as the field around a bar magnet. As long as the current flows, the coil acts like a bar magnet.

Cosmic rays causing ionising radiation and **solar flares** from the Sun also interfere with the operation of **artificial satellites**.

Cosmic rays:

- are fast-moving particles which create gamma rays when they hit the atmosphere
- spiral around the Earth's magnetic field to the poles
- cause the Aurora Borealis.

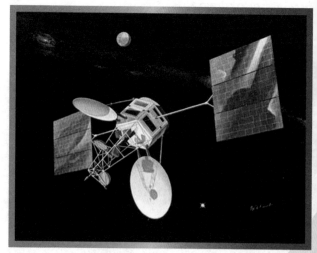

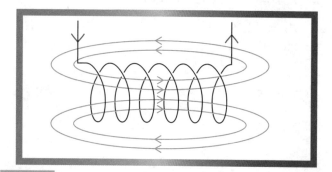

Solar flares:

- are clouds of charged particles from the Sun

- are ejected at high speed

- produce strong disturbed magnetic fields.

Artificial satellites are used for:

- telecommunications

- weather prediction

- spying

- satellite navigation systems.

👁 PROGRESS CHECK

1. What types of waves do mobile phones use?

2. Give **two** ways that signal loss can be reduced.

3. Can living near a phone mast be dangerous?

4. State **two** sources of artificial satellite interference.

5. Suggest **two** uses for artificial satellites.

❓ EXAM QUESTION

The following diagram shows a coil of wire that has a current passing through it.

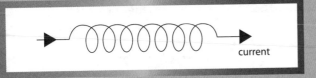

current

a. Sketch the shape of the magnetic field around the coil of wire.

b. What does the magnetic field around the Earth protect us from?

c. How does the Earth's magnetic field affect cosmic rays?

d. Name one other source of magnetic fields in space.

Answers

Day 1

pages 4–5
How Science Works
PROGRESS CHECK
1. The variable we choose to change in an experiment
2. The variable that we measure in an experiment
3. A variable that can be put in order
4. A variable that can have any whole-number value
5. Close to the true value

EXAM QUESTION
a. The force applied
b. The length of the spring
c. Using the same spring, etc.

pages 6–7
Motion
PROGRESS CHECK
1. Because they need to measure the distance travelled in a certain time
2. Average velocity = displacement / time
3. 20 m/s
4. Speed has magnitude (size) only, it is a **scalar** quantity. Velocity has magnitude and direction, it is a **vector** quantity.
5. Yes, because the direction changes.

EXAM QUESTION
1. Acceleration = change in velocity / time taken
2. 5.0 m/s/s
3. 10 m/s/s
4. The velocity at two different points and the time between them

pages 8–9
Calculating Motion
PROGRESS CHECK
1. 5 m/s
2. 21 m/s
3. 3 m/s north
4. 5.4 m/s
5. 5 N

EXAM QUESTION
a. 0 m/s   b. 35 m/s
c. 61 m   d. 7 m

pages 10–11
Graphs for Motion
PROGRESS CHECK
1. The velocity
2. Greater velocity
3. The acceleration
4. The area under the graph
5. Negative acceleration (deceleration)

EXAM QUESTION

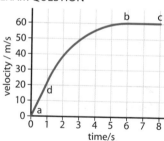

pages 12–13
When Forces Combine
PROGRESS CHECK
1. A free body diagram shows the forces that act on a single body
2.

3. 400 N downwards
4. Because he pushes backwards on the boat in order to move forwards
5. Action – the man pushes backwards on the boat with a contact force. Reaction, the boat pushes forwards on the man with a contact force.

EXAM QUESTION
a. No
b. They are not the same kind of force; they also act on the same body
c. The contact force of the girl downwards on the bench (caused by the gravitational force of her weight)
d. A free body force diagram

pages 14–15
Forces and Motion
PROGRESS CHECK
1. 10 m/s/s
2. It would increase
3. A parachutist
4. Towards the centre
5. Increasing the mass of the body, the speed of the body or decreasing the radius of the circle

EXAM QUESTION
a. and b.

c. When the magnitude of the drag/air resistance equals the weight
d. Terminal velocity

pages 16–17
Momentum and Stopping
PROGRESS CHECK
1. 750 kgm/s
2. When a force acts on a body that is moving, or able to move, a change in momentum occurs
3. Where part of a car is designed to collapse steadily in a collision spreading stopping over longer time
4. The time it takes for a vehicle to stop
5. Greater stopping time means that less force is exerted on the passengers

EXAM QUESTION
a. 6.0 kgm/s　b.　3.0 kg
c. 6.0 kgm/s　d.　2.0 m/s

Day 2

pages 18–19
Safe Driving
PROGRESS CHECK
1. The distance travelled between the need for braking occurring and the brakes starting to act
2. The distance taken to stop once the brakes have been applied
3. Driver tiredness; distractions; lack of concentration (any two)
4. If brakes are applied too hard the wheels will lock (stop turning) and the car will begin to skid. ABS braking systems detect this locking and automatically adjust the braking force to prevent skidding
5. Electric windows, cruise control, adjustable seating

EXAM QUESTION
a. 50 m　b. 10 m
c. 5 m　d. Wet roads, speed

pages 20–21
Energy and Work
PROGRESS CHECK
1. Joules
2. kinetic energy = ½ × mass × velocity² / KE = ½ mv²

3. 9 J
4. Weight = mass × gravitational field strength
5. 45 N

EXAM QUESTION
a. Gravitational potential energy
b. 30 000 J
c. 30 000 J
d. 24 m/s

pages 22–23
Work and Power
PROGRESS CHECK
1. Work done is equal to energy transferred; work is done when a force moves an object
2. Joules
3. Climbing stairs, pulling a sledge (any reasonable answer)
4. 100 J
5. For an object that is able to recover its original shape, elastic potential is the energy stored in the object when work is done to change its shape

EXAM QUESTION
a. Gravity (his weight)
b. It is the distance moved in the direction of the force
c. 1000 J
d. No, he does the same

pages 24–25
Static Electricity
PROGRESS CHECK
1. A material that charge (electrons) can pass through
2. A material that charge (electrons) cannot pass through
3. Aircraft fuel causes static charge as the fuel rubs against the pipe. A spark could ignite the fuel vapour. The aircraft and the tanker are earthed to avoid the charge building up
4. Correct earthing or use of insulating mats
5. Synthetic clothing clings; dirt and dust are attracted to TV screens, etc; static shocks from

cars, earthed conductors, etc. (any one)

EXAM QUESTION
a. Because each droplet repels the others, so they spread out evenly
b. No, the opposite charge
c. An even coat and less waste
d. In photocopiers

pages 26–27
Electricity on the Move
PROGRESS CHECK
1. Current is measured in Amps or Amperes, it is equal to the amount of charge that flows every second
2. 5.0 C per second
3. a. True
   b. False, they flow from negative to positive
   c. False, ions do not move at all
   d. True

EXAM QUESTION
a., b. and c.

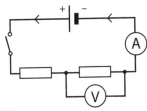

d. The current will stop flowing

pages 28–29
Circuits
PROGRESS CHECK
1. 1.5 Ω
2. It will decrease
3.

4. 4.1 A
5. It would decrease

EXAM QUESTION
a. 10 Ω
b. 1.2 A
c. 6.0 V
d. In parallel

Day 3

pages 30–31
Electrical Power
PROGRESS CHECK
1. 1.5 W
2. 240 000 J
3. 0.067 kWhs
4. 4.8 A
5. 96p

EXAM QUESTION
a. 144 000 J, 0.04 kWhs
b. 28.8p
c. 0.005 kWhs
d. 3.6p

pages 32–33
Electricity at Home
PROGRESS CHECK
1. The resistance of a filament lamp increases with more current, the lamp gets hotter.
2.

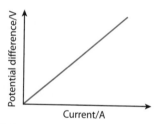

3. It increases
4. Fire alarm
5. 5.0 A

EXAM QUESTION
1. The fuse
2. The earth wire connects the metal parts of the hair dryer to earth and stops the device becoming dangerous to touch if there is a loose wire. If a large current flows to earth, the fuse melts and breaks the circuit
3. A residual current circuit breaker improves the safety of a device by disconnecting a circuit automatically whenever it detects that the flow of current is too high

4. The hair dryer is insulated with a plastic case and the wires themselves are also insulated

pages 34–35
Controlling Voltage
PROGRESS CHECK
1.

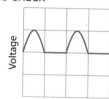

2. Half wave rectification
3. Full wave rectification
4.

5. To smooth out the current

EXAM QUESTION
1. Resistance is high in the dark and low in the light
2.

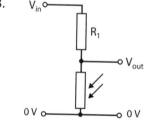

3. V_{in}

R_1

V_{out}

0 V   0 V

4. To switch on an exterior light

pages 36–37
Electronic Applications
PROGRESS CHECK
1.

IN — NOT — OUT

2.

Input	Output
1	0
0	1

3.

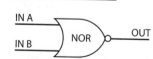

IN A
IN B — NOR — OUT

4.

Input A	Input B	Output
0	0	1
0	1	0
1	0	0
1	1	0

5. To switch on the interior light in a car when either the driver's door or the passenger's door or both doors are open

EXAM QUESTION
a. The fire alarm will sound
b. The fire alarm will sound
c. The fire alarm will sound and the sprinklers will come on
d.

smoke	heat	alarm	sprinklers
0	0	0	0
0	1	1	0
1	0	1	0
1	1	1	1

pages 38–39
Producing Electricity
PROGRESS CHECK
1. A device to measure current
2. In series (next to) a component
3. It is more efficient, less energy wasted as heat
4. Up to 400 000 V
5. They do not spoil the landscape

EXAM QUESTION
a. Increasing the strength of the magnet, increasing the number of turns in the coil of wire, increasing the speed of rotation of the magnet
b. AC
c. DC is only in one direction, AC continuously changes direction
d. In series with the lamp

pages 40–41
Motors and Transformers
PROGRESS CHECK
1. 19 V
2. Increasing the number of turns on the coil, increasing the current, using a stronger magnet
3. field

. current
. force (thrust)

EXAM QUESTION
a. 10:1
b. Primary on left
c. Step down
d. At 'output voltage'

pages 42–43
Electricity in the World
PROGRESS CHECK
1. Increased the processing speed
2. Sending many signals at once
3. It has very little friction
4. The decreasing of resistance to almost zero in certain materials at extremely low temperatures
5. An analogue signal is continuously variable, a digital signal is either on or off

EXAM QUESTION
1. The waves are all the same frequency and in phase
2. CD player
3. Touchtone dialing

Day 4

pages 44–45
Harnessing Energy 1
PROGRESS CHECK
1. Chemical energy to heat to kinetic energy (in a turbine which turns a generator) to electrical energy
2. Solar, wind power and moving water
3. No pollution
4. It can disturb habitats
5. Heat to kinetic to electrical

EXAM QUESTION
a. Renewable
b. It requires no fuel, it causes no pollution
c. Depends on wind speed, can disturb habitats
d. Kinetic energy of the wind to electrical energy

pages 46–47
Harnessing Energy 2
PROGRESS CHECK
1. Requires large areas of land
2. Sulfur dioxide
3. Carbon dioxide
4. CFCs
5. It reflects radiation from the city, causing warming

EXAM QUESTION
a. Efficiency =
$$\frac{\text{useful energy output}}{\text{total energy input}} \times 100\%$$
b. 25%
c. It is lost as heat
d. They have cooling systems

pages 48–49
Keeping Warm
PROGRESS CHECK
1. Convection, conduction and radiation
2. Conduction
3. It expands and becomes less dense
4. Matt black
5. Loft insulation, double glazing, curtains

EXAM QUESTION
a. Conduction
b.

warm inside — cold outside — cavity — bricks

c. By filling the cavity with insulation
d. Cool

pages 50–51
Heat
PROGRESS CHECK
1. To absorb more energy
2. A measurement of hotness and is measured in degrees celsius (°C)
3. Heat is a measurement of energy, measured in joules

4. They have a large surface area compared with their mass so lose heat quickly
5. Tiles are good thermal conductors and transfer heat away quickly

EXAM QUESTION
a. They are knocked loose from the silicone crystals
b. Surface area and light intensity
c. They can operate in remote locations
d. They are expensive

pages 52–53
Heat Calculations
PROGRESS CHECK
1. How much energy 1 kg of a material needs to increase (or loses when decreasing) its temperature by 1 °C. The unit of specific heat capacity is Joules per kg per degree
2. At normal pressure water freezes and melts at 0 °C
3. The specific latent heat of a substance is the amount of energy required to change the state of 1 kg of that substance at constant temperature
4. 42 000 J
5. 1925 J

EXAM QUESTION
a. The water needs energy to change state at constant temperature
b. 420 000 J
c. Some heat will be lost to the surroundings
d. 2 260 000 J

pages 54–55
Temperature
PROGRESS CHECK
1. Temperature is a measure of hotness in degrees
2. Heat is a measure of energy in Joules
3. Absolute zero

4. They stop moving
5. 0 K

EXAM QUESTION
1. Pressure is proportional to temperature in degrees Kelvin
2. Hotter particles move faster and bombard the surface more frequently and with greater force
3. P/T = constant
4. 240,000 Pa

Day 5

pages 56–57
Inside the Atom

PROGRESS CHECK
1. No charge, neutral
2. In an unstable nucleus a neutron may spontaneously change into a proton. When it does this, it emits an electron. This is called beta radiation.
3. Huge distances
4. The ability of the radiation to cause other particles to gain or to lose electrons
5. Strong

EXAM QUESTION
1. a. Isotopes
    b. 146
    c. A Helium nucleus (2 protons and 2 neutrons)
    d. $^{238}_{92}U \rightarrow \, ^{234}_{90}Th + \, ^{4}_{2}\alpha$

pages 58–59
Smaller Particles

PROGRESS CHECK
1. A similar particle to an electron with a positive charge
2. Quarks, electrons
3. When electrons are 'boiled off' hot metal filaments
4. TV tubes, oscilloscopes, the production of X-rays
5. By an electric field created by parallel charged metal plates

EXAM QUESTION

a.

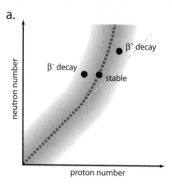

pages 60–61
Radioactivity

PROGRESS CHECK
1. In fire alarms
2. To find leaks or blockages in underground pipes
3. To limit the exposure and therefore the damage to non-cancerous tissue
4. They are produced in different ways
5. X-rays are easier to control.

EXAM QUESTION
1. The time it takes for the number of undecayed nuclei to halve
2. 3 days
3. 50 counts per second
4. The date of artefacts

pages 62–63
Safe Radiation

PROGRESS CHECK
1. Rocks, cosmic rays and human use (e.g. medical)
2. Some media reports claim that it is
3. Because it can be breathed into the body
4. No
5. Yes

EXAM QUESTION
1. Fission
2. Risk of accidental emission of radioactive material; waste material disposal
3. The Earth's atmosphere and magnetic field and the ozone

layer protects it
4. Use tongs or gloves, minimise the exposure time

pages 64–65
Radiation and Science

PROGRESS CHECK
1. If a model is successfully tested, scientists are more likely to accept it
2. Relativity
3. The electron
4. Because the electrons look a little bit like plums in a pudding
5. Electrons can behave like clouds of charge, or even waves

EXAM QUESTION
a. Alpha particles
b. Because it is very thin
c. They passed straight through the gold foil
d. Most of an atom is empty space, most of the mass of an atom is compressed into a tiny volume in the centre called a nucleus, the nucleus of an atom has an overall positive charge

pages 66–67
Science and Medicine

PROGRESS CHECK
1. The amount of oxygen in the blood
2. Stress, age, body mass
3.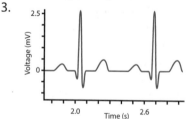

4. They annihilate one another and produce a gamma ray
5. Some people believe that being highly sedated is unethical

EXAM QUESTION
1. Positron emission tomography
2. To minimise damage to the

patient's cells

3. To find cancer tumours

4. Due to the short half-lives of the radioisotopes used, they must be created in a cyclotron in the hospital

pages 68–69
Nuclear Power

PROGRESS CHECK

1. Two lighter nuclei and two or three neutrons
2. The two lighter nuclei
3. It can be encased in glass or concrete and buried.
4. Thousands of years
5. The joining of two atomic nuclei to form a larger one

EXAM QUESTION

a. It slows down the neutrons (usually graphite or water)
b. They absorb some of the neutrons
c. No greenhouse gases; small amounts of fuel give large amounts of energy
d. High maintenance costs; risk of accidental emission of radioactive material

Day 6

pages 70–71
Waves

PROGRESS CHECK

1. Wavelength is the distance between any point on a wave and the same point on the next wave
2. Amplitude is the maximum displacement of the waves from rest position
3. The total time taken for a single wave to pass a point
4. Longitudinal
5. Transverse

EXAM QUESTION

a. P-waves and S-waves
b. P-waves

c. P-waves

d. Only P-waves can travel through liquid, so when S-waves are stopped by parts of the core, we know that these parts must be liquid

pages 72–73
Reflection

PROGRESS CHECK

1. Angle of incidence = angle of reflection
2. Virtual, upright, same size as the object
3. A constructed line perpendicular to any reflecting surface at the point where the incident ray hits the surface
4. Shaving mirror, makeup mirror
5. Driving mirror

EXAM QUESTION

a., b. and c.

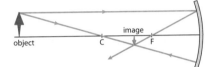

(not to scale)

d. Real, inverted, smaller than the object

pages 74–75
Wave Behaviour 1

PROGRESS CHECK

1. Lines drawn one wavelength apart, perpendicular to the direction of the waves
2. Areas where the waves add together (reinforcement)
3. An even number of half wavelengths (i.e. a whole number of waves)
4. No
5. Sunglasses / camera lens

EXAM QUESTION

1. The spreading out of waves when they pass through a gap
2. It stays the same

3.

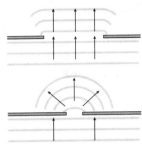

4. The size of the gap must be similar to the wavelength for diffraction to be significant

pages 76–77
Wave Behaviour 2

PROGRESS CHECK

1. When a wave changes direction as it travels from one medium to another
2. When it leaves water and enters air
3. Because of apparent depth due to refraction
4. When a narrow beam of white light passes through a prism it separates into a spectrum of colour due to refraction
5. Sending many signals along an optical fibre at once

EXAM QUESTION

a. Total internal reflection

b.

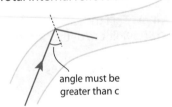

angle must be greater than c

c. About 41°

d.

pages 78–79
Lenses
PROGRESS CHECK
1. Converging lens
2. The distance from the optical centre of the lens to the focal point
3. Real, inverted, the same size as the object
4. A real image can be focused on a screen, a virtual image cannot
5. Upright, virtual

EXAM QUESTION
a., b., c. and d.

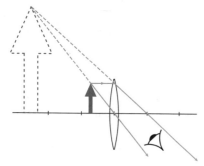

Day 7
pages 80–81
The Electromagnetic Spectrum
PROGRESS CHECK
1. The type of radiation and the size of the dose
2. It may kill the cells
3. Ultraviolet to detect forged bank notes
4. Optical waves for iris recognition
5. Infrared to monitor temperature

EXAM QUESTION
1. Microwaves, gamma rays, radio waves, visible light
2. 10^{-15} m
3. 1 mm
4. Microwaves are absorbed by particles on the outside layers of food to a depth of about 1cm, increasing the kinetic energy of the particles. The heat is transferred to the centre by

conduction or convection

pages 82–83
Sound
PROGRESS CHECK
1. No, longitudinal
2. Ultrasound
3. Larger amplitude makes a louder sound
4. Higher frequency makes a higher pitch/note

EXAM QUESTION
a. 30 000 Hz
b. The time taken for reflections to reach a detector
c. Pre-natal scanning, breaking down kidney stones, measuring speed of blood flow
d. It can produce images of soft tissues, does not damage living cells

pages 84–85
Beyond our Planet
PROGRESS CHECK
1. Every 24 hours
2. The Milky Way
3. A star that has expanded outer layers near the end of its life
4. It becomes a red giant, then explodes as a supernova, then becomes a very dense neutron star or a black hole
5. Because of the very strong gravity

EXAM QUESTION
a. Comets and asteroids
b. Because the large gravity of Jupiter disrupts the formation of a planet
c. Craters, sunlight blocked by dust, climate change
d. An object that orbits a planet

pages 86–87
Space Exploration
PROGRESS CHECK
1. It can cause deterioration of bones and heart
2. Artificial gravity, air supply,

enough food and water
3. The distance light travels in a year
4. Temperature, magnetic field, atmosphere
5. The expanding universe and cosmic rays

EXAM QUESTION
1. Mass is the amount of matter something is made of and is measured in kilograms. The weight of a body is a measure of the amount of gravitational force acting on it and is measured in Newtons
2. 73.5 N
3. The mass does not change, the weight decreases
4. Backwards

pages 88–89
Communication
PROGRESS CHECK
1. Microwaves and radio waves
2. Limiting the distance between transmitters, positioning transmitters as high as possible
3. There have been claims that it is
4. Cosmic rays causing ionising radiation and solar flares from the sun
5. Spying, satellite navigation systems

EXAM QUESTION
a.

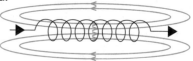

b. Ionising radiation from space
c. They spiral around the magnetic field to the poles, causing Aurora Borealis
d. Solar flares